By Norman Cousins

THE IMPROBABLE TRIUMVIRATE

John F. Kennedy

Pope John

Nikita Khrushchev

W · W · NORTON & COMPANY

New York · London

Copyright © 1972 by Norman Cousins. All rights reserved.
Published simultaneously in Canada by Stoddart, a subsidiary of
General Publishing Co. Ltd., Don Mills, Ontario.

First published as a Norton paperback 1984

Library of Congress Cataloging in Publication Data
Cousins, Norman.
The improbable triumvirate
1. World politics—1955–1965.   2. United States.
Treaties etc., 1961–1963 (Kennedy) Treaty banning
nuclear weapon tests in the atmosphere, in outer space
and under water. 3. Kennedy, John Fitzgerald, Pres.
U.S., 1917–1963. 4. Joannes XXIII, Pope, 1881–1963.
5. Khrushchev, Nikita Sergeevich, 1894–1971.
I. Title.
D843.C63   1972   327'.11'09046   72-4819

ISBN 0-393-30162-1

W. W. Norton & Company, Inc.
500 Fifth Avenue, New York, N.Y. 10110
W. W. Norton & Company Ltd.
37 Great Russell Street, London, WC1B 3NU

1 2 3 4 5 6 7 8 9 0

CONTENTS

Photographs appear following page 64

*I*T WAS a long time ago and he may have forgotten it, but this book was originally suggested by William Epstein, director of the Disarmament Affairs Division of the United Nations. Senator Robert F. Kennedy encouraged the author in the writing of this book and supplied background data on various aspects of the Cuban crisis. Pierre Salinger, press secretary to President John F. Kennedy, and Theodore Sorensen, special assistant to President Kennedy, filled in details on the private exchanges between President Kennedy and Chairman Khrushchev. This grateful acknowledgment in no way commits any of the foregoing men to approval of the treatment of their material or of the conclusions of this book.

My warm thanks to Dana Little and Deborah Appel, then on the staff of *Saturday Review*, and to Sharon Fass, formerly of *Saturday Review* and now of *World Magazine*. They provided research assistance, brought order out of mounds of data, retyped the manuscript perhaps a dozen times or more, and gave advice on editing.

Portions of this manuscript appeared originally in the *Saturday Review*.

N.C.

An Asterisk to History: Some Footnotes on the Relationship of John F. Kennedy, Pope John XXIII, and Nikita S. Khrushchev

*B*EGINNING LATE IN 1962, and for most of the next year, a new spirit of hopefulness was abroad in the world. The upturn was largely the result of a strange and magnificent interaction among three men: President John F. Kennedy, Pope John XXIII, and Prime Minister Nikita Khrushchev. Ideologically and personally, they were one of history's most implausible triumvirates: an American President, a Pope, a Communist. What brought them together was the vulnerability of civilization to modern destructive power. The particular event that gave blistering reality to their common concern was the Cuban crisis of October, 1962.

The crisis over Cuba was bridged but the terror it produced created a new sense of resolution and dedication. At American University on June 10, 1963, President Kennedy proposed an end to the Cold War. The proposal produced an affirmative response from Prime Minister Khrushchev. Pope John used the full weight of the Papacy to speak about the need for world peace under law. His encyclical *Pacem in Terris* was perhaps his single most important document. It called for a new spirit of world cooperation in building an enduring world order.

The three men had a profound respect for one another; all understood the extent to which their combined role was historically necessary, however diverse or contradictory their backgrounds. For a brief period their candle burned brightly. Then very quickly, the trio was lost to history. President Kennedy was assassinated, Pope John died of cancer, and Nikita Khrushchev was replaced in office.

The account I give here of some incidents bearing on that relationship is hardly more than an asterisk to history, but it does touch upon some remarkable exchanges among the trio. Through a strange combination of circumstances, I found myself an emissary for Pope John to the Kremlin. An off-shoot of this mission involved President Kennedy.

A starting point for the personal aspects of this story is early March, 1962, when Father Felix P. Morlion, O.P., president of Pro Deo University in Rome, visited me in New York. We spent three hours at my office, then went out to dinner. Father Morlion was a large, hearty man who had a gift for laughter and a genius for making people feel at ease. It quickly became clear that Father Morlion was one of the most versatile intelligences I had ever met. His range included an intimate knowledge of American history and institutions, world politics, modern science and technology, motion picture techniques, and European literature. He was a native of Belgium but was equally at home in France, Italy, Germany, or the United States, the languages of which he spoke with astonishing facility. I learned that, during World War II, he had worked in the anti-Nazi underground; he had a price on his head because of his activities in harboring Jews. His philosophical outlook was that of world citizen. Padre Morlion's universalism was reflected in the basic intellectual environment of Pro Deo University, which he founded in 1945 with the help of the then Deputy Secretary of State for Ordinary

Affairs of the Vatican, Giovanni Battista Montini, who later became Pope Paul VI.

At that first meeting he said he had come to the United States for the purpose of attracting support for Pro Deo University, which from its start had been interreligious and international, a fact he attributed in large part to the influence of Pope John XXIII and Cardinal Montini. Father Morlion spoke animatedly of his hopes for the development of Pro Deo University as a training center for world citizenship, bridging gaps not just between East and West but within the West itself; gaps, too, between theologian and scientist, philosopher and activist.

The Padre said he was also working on a book about Pope John. At that time the full force of Pope John's ideas had not yet broken on the world. His belief in the principles of fraternalism and world understanding, of course, was well known, and the warmth of his personality was beginning to be widely felt; but it was only some months later, following the encyclical *Pacem in Terris,* that Pope John came full size, historically. Hence, when Father Morlion began to talk about the Pope's ideas and hopes, I could feel the excitement of profound changes in the making.

"You must believe me," Father Morlion said, "when I tell you that dramatic ideas are shaping up and all the world will come to acclaim and love this gentle man, Pope John. He is not arbitrary or fixed. He has a profound respect for people of all faiths. He wants to help save the peace."

Father Morlion said Pope John wanted to be useful and relevant but there was much uncertainty about how this might best be done. He asked whether I thought the American people would regard strong statements by the Pope about the need for bold measures in behalf of world peace as an unwelcome intrusion into political affairs. I told the Padre I had no way of knowing how Americans, or any other people, would react, but that it was difficult for me to believe there was anything unnatural about a strong call for peace by any of the world's great religious leaders. If anything, religion had yet to perceive and act adequately upon the spiritual connotations of world breakdown leading to war. From the theological viewpoint, nuclear war was not just a war of nation against nation, or even man against man—it was a war against God. Man was now on the verge of smashing at the conditions of life and not just at life itself. Nuclear war could alter man's genetic structure; it could disfigure his life and create a deformed environment. If concern over such facts was not a matter for Papal intervention, it was hard to determine what was.

The Padre agreed that great initiatives by the Vatican for peace were neeeded. He said he believed the Pope was encouraged by the growth of anti-war forces inside the Soviet Union. These long-awaited opening-up trends inside Russia were profoundly significant and ought to be explored and tested. Father Morlion added that any appeal for peace by Pope John had to be carefully prepared. More particularly, it was necessary to develop access to the Russian leaders so that the Pope's appeal would carry weight.

He said Pope John believed world understanding might be advanced through an International Cooperation Year similar to the highly successful International Geophysical Year. (At that time—summer, 1962—Prime Minister Nehru had not yet made his formal proposal to the United Nations for an International Cooperation Year. So far as is known, there was no connection between the Pope's proposal and Mr. Nehru's.)

As fall approached, we continued to discuss these possibilities. Father Morlion asked about the Dartmouth conferences. These conferences, bringing together academicians, writers, and scientists from the United States and U.S.S.R., began at Dartmouth College in Hanover, New Hampshire, in October, 1960, and were continued in the Crimea in June, 1961. A third meeting was scheduled at Phillips Academy, Andover, Massachusetts, for October, 1962. The purpose of the conferences was to explore problems facing the statesmen of both nations in an effort to see whether initial approaches might be defined which could serve as the basis for appropriate action by government leaders in both countries. Father Morlion was ineligible to be a participant or observer at Andover, but he asked whether he might find an opening at the conference to talk to the Russians.

The Americans and Russians began their week-long conference at Andover on October 21, 1962. For at least five days before they met, tension had been building up over reports of Russian strategic military shipments to Cuba. This tension was reflected at the get-acquainted dinner Sunday evening at Andover, when both groups gathered around a television set to hear President Kennedy's talk to the nation. The President confirmed the existence of Soviet missiles in Cuba. He said the situation was intolerable and that, with full realization of the risks, he had ordered American naval vessels to intercept further Soviet military shipping to Cuba.

I watched the American and Russian delegates as they listened to the President. Among the Americans were Philip E. Mosely, director of studies, Council on Foreign Relations; John B. Oakes, editor of editorial page, *The New York*

Times; Robert B. Meyner, governor of New Jersey, 1954–1962; Arthur Larson, director of World Rule of Law Center, Duke University; Thomas B. Coughran, executive vice president of Bank of America; Shepard Stone, director of International Affairs Program, Ford Foundation; Paul M. Doty, member of President's Science Advisory Committee; Lester B. Granger, president of International Conference of Social Work; Margaret Mead, anthropologist; Louis B. Sohn, Bemis professor of international law, Harvard University; Herman Steinkraus, former chairman of Bridgeport Brass Company; Harry Culbreth, vice president of Nationwide Insurance Company; Norris Houghton, co-managing director of Phoenix Theatre, New York.

Among the Soviet delegates were Academician Evgeny K. Fedorov, Chief Scientific Secretary, U.S.S.R. Academy of Sciences; General Nikolai Talensky, military theoretician; Georgii (Yuri) Zhukov, former chief of Cultural Exchange Program and leading Soviet journalist; Boris Nikolaevich Polovoi, writer, editor, deputy, Supreme Soviet, R.F.S.F.R.; Grigory Shumeiko, journalist and member of editorial board, Soviet Trade Unions; Dr. Vladimir Kovanov, surgeon and corresponding member of U.S.S.R Academy of Medical Sciences; Professor Alla Massevich, astronomer.

Both Americans and Russians wore the same expression of acute concern as they listened to the President. After it was over the Russians stood up. Their chairman, Evgeny K. Fedorov, asked if they could be excused for a few minutes to meet by themselves. The Americans recognized, of course, that the Russians were confronted by a special situation. If war broke out they would be interned, assuming, of course, that the war lasted long enough to make this a problem. We felt it necessary to discuss frankly with our Russian guests their own feelings about arranging for their immediate return to Moscow.

For our part, we thought it important not to give our guests synthetic reassurances about the crisis. Consequently, when both groups reconvened at about 9:00 P.M. that night, I, as co-chairman of the American delegation, opened the meeting by saying that the Americans recognized that the sudden crisis made it necessary to consider the advisability of adjourning the meeting.

"We have considered this matter," Dr. Fedorov replied. "Gentlemen, we are in your hands. We will do whatever you wish us to do. If you wish to proceed with the conference, we will stay. If not, we will leave."

I didn't think it was necessary to have a private caucus of the American group in order to answer Dr. Fedorov's question. Moreover, I thought it might be salutary for us to decide openly. I asked for a show of hands of those Americans who wished to proceed with the conference as originally planned. Instantly, and without a single exception, the Americans voted affirmatively.

"Very well. We will go on with the conference," Fedorov replied.

He asked for the floor, and for the next half-hour painstakingly defended the Soviet action in Cuba, characterizing the President's decision to blockade Soviet ships as arrogant, aggressive, illegal. The rebuttal from the American side was no less explicit or comprehensive. We asked General Talensky whether it was the Russian intention to stay in Cuba indefinitely. Speaking without acrimony and with great earnestness, General Talensky explained that the Russian purpose was not to convert Cuba into a Soviet military base. He said the missiles were intended for Cuban defense purposes in response to the clamor in the United States for an American invasion.

"Then it is the purpose of the Soviet Union to turn the missiles over to Castro?" Phil Mosely asked.

"Yes," said General Talensky.

"In that case," Phil Mosely said quietly, "I ask you whether the cause of peace, which is what concerns all of us here, will be advanced with missiles in the hands of Castro. Does anyone here suppose for a moment that Castro's possession of missiles this close to the United States would not at some point trigger a war?"

Yuri Zhukov, one of the editors of *Pravda,* asked whether an American invasion of Cuba might not also trigger a war.

The debate at Andover that week was strenuous, sometimes strident, but two things became clear as it spilled over into the second day. One was that the Cuban crisis didn't interfere with the cordiality of the Russians or their desire to have a productive conference. The second was that both Russians and Americans, as private citizens, showed a clear desire to find a way out of the crisis.

It was at this point that one of the advantages of unofficial meetings such as ours became apparent. As private citizens we didn't have to argue from or to fixed positions. It was possible to explore alternatives without making commitments. These alternatives could be scrutinized by our respective governments, which saw in the Andover meeting a chance to put out feelers on proposals that might break the deadlock.

At about this time Father Morlion arrived at Andover. In the light of our earlier discussions about the appropriateness of the Pope's possible activity in behalf of world peace, Father Morlion asked whether Papal intervention in the Cuban crisis —even if only in the form of an appeal for greater responsibility—might not serve an important purpose. It was quite possible that both the United States and the U.S.S.R. might be in a better position to react favorably to an outside proposal, whereas the same proposal made by either party directly might be rejected by the other automatically without regard to merit. With the encouragement of those members of both delegations whom he took into his confidence, Morlion telephoned the Vatican. A few hours later he came back with word that the Pope was intensely apprehensive about the Cuban crisis and wanted to help avert a hideous culmination. First, however, the Pope wished to be certain his moral intervention would be acceptable. Would a proposal to both nations be acceptable, Father Morlion asked, that called for a withdrawal both of military shipping and the blockade?

I telephoned the White House and spoke to Ted Sorensen, who called back after conferring with the President. He said the President welcomed the offer of Pope John's intervention and, indeed, welcomed any initiatives that would prevent an escalation of the Cuban crisis. But the President could not encourage Pope John to believe that his proposal met the central issue. That issue was not so much the shipping but the presence of Russian missiles on Cuban soil. Those missiles had to be removed—and soon—if the consequences of the crisis were to be averted.

I relayed the information to Father Morlion, who telephoned the Vatican. Father Morlion also consulted with the leaders of the Soviet delegation at Andover, one of whom telephoned Moscow and reported that the Pope's proposal calling for withdrawal both of the military shipping and the blockade was completely acceptable to Premier Khrushchev.

The next day Pope John issued his call for moral responsibility in the Cuban crisis. In line with President Kennedy's reservations, he made no specific reference to the military shipments or the blockade. Instead, he directed himself to the clear obligation of political leaders to avoid taking those steps that could lead to a holocaust. He said that not just the Americans and Russians but all the world's peoples were involved, and that their fate could not be disregarded. He said that history would praise any statesman who put the cause of mankind above national considerations.

The Pope's appeal made headlines throughout the world, including the Soviet Union. Any impact it may have had on the contending governments, however, was not apparent; the crisis continued to deepen.

By the time our conference concluded at the end of the week, there was a sense of a fast-approaching saturation of tension. Even though the conferees had discussed all the matters on the prepared agenda having to do with the need for increased understanding of each other's position, the one issue on everyone's mind was Cuba.

Enroute to New York by chartered bus, the Russian and American delegates stopped at my home in New Canaan for a snack. While we were going into the house, someone heard a radio news report saying Premier Khrushchev announced he was removing the missiles from Cuba and had written a long letter to President Kennedy expressing the hope that the lessons learned during this crisis could be profitably turned to the promotion of peace.

There was nothing restrained in the toasts of the Russians as they prepared to bid farewell to their American hosts. The response of the Americans was equal to the occasion.

Several days later, Yuri Zkukov went to Washington to meet with Pierre Salinger, President Kennedy's press secretary. The discussions were wide-ranging, the emphasis being on specific measures that might be taken by both countries to improve their relations and to ease world tensions. The report we received from Mr. Salinger on these meetings was positive. We could ponder the contrast between the mood of the Dartmouth Conference the night it began at Andover and the general mood of hopefulness that followed the resolution of the Cuban crisis. But that auspicious new mood was not to last very long.

*I*t was against the background of the Pope's message on Cuba that Father Morlion informally explored with some of the Soviet delegates the possibility of further communication between Rome and Moscow in the cause of a workable peace. He said he knew how implausible this sounded, given the historical incompatibility between these two groupings, but humankind was now faced with overriding needs. He told the Russians he had reason to believe that I would be acceptable to the Vatican for the purpose of undertaking preliminary contacts between Rome and Moscow, and he asked if I would be equally acceptable to the Russians. This approach was consistent with Pope John's determination, in light of the horror of the Cuban crisis, to do what he could toward helping to free the world's peoples from the threat of a nuclear holocaust.

Father Morlion expressed the view to the Soviet citizens that private contacts between the Vatican and Moscow might lead to important understandings. In particular, he proposed that an individual—unofficial and unattached—who was acceptable to both parties might initiate an exchange of ideas.

Father Morlion emphasized he was speaking entirely as an individual in making this proposal; but he felt individual citizens had the responsibility to undertake initiatives which might not always be feasible or possible for officials. If the initiatives worked out well, the officials could appraise the results and follow through. If the initiatives were unproductive or unworkable, they could be dropped. In any case, the leaders could remain uncommitted.

The Soviet delegates said they would make inquiries on all these points after they returned to Moscow, and would reply by letter or cable.

For several weeks after the Russians left, Father Morlion would telephone me each day at the *Saturday Review* to find out whether word had arrived from Moscow about the project. Meanwhile, the sense of hope that had sprung up at the end of the Cuban crisis week began to fade. Specific areas of possible accord—a ban on nuclear testing, Berlin, and outer space—seemed even more remote than before Cuba. A new downward drift seemed to be setting in.

One day late in November I received a telephone call from Ambassador Anatoly F. Dobrynin in Washington. He said the project proposed by Father Morlion at Andover had been approved and that December 14 was suggested as a possible date for a visit by me to Premier Khrushchev in Moscow on behalf of the Vatican.

Under United States law American citizens are forbidden to hold discussions with heads of governments on matters that could have a bearing on the policies of those countries toward the United States. Although the United States was not directly involved in my private mission, I thought it best to inform the government. I went to Washington and discussed the project with Pierre Salinger. Two days later he telephoned me in New York to say he had spoken to the President and that there were no objections to the trip. Salinger said that Ralph

Dungan, a presidential assistant, had been assigned by the President to follow the project and keep him informed. The President also felt he ought to speak to me just before my departure.

A week later I had a long session with Ralph Dungan, who, like his chief, represented a type in American politics that was familiar at the time of the Philadelphia Constitutional Convention—men of ideas, young, resourceful, confident without being cocksure.

President Kennedy entrusted Dungan with a wide variety of responsibilities, all the way from processing political appointments to dealing with religious matters. In addition, Latin America was considered his special province. Deceptively relaxed and easygoing in manner, Ralph Dungan worked under unbelievably severe pressures. He was a superb listener, and had a faculty for total recall of important conversations. Like most of the other New Frontiersmen, he was vigorous, highly intellectual, devoted to his chief. I found him extremely well informed on Vatican affairs, especially on the intricate question of Church-state relationships.

Several days later I was invited to the White House. I went through the front gate at the White House, where my name was checked off the guard's list. Then I was escorted into the Cabinet room on the ground floor; it looked out on a garden enclave and adjoined the portico connecting the living quarters of the White House with the executive offices. Beyond the enclave was the White House lawn, which seemed far less manicured and level than it appeared from the street. I observed several well-sheltered nooks, one of which was modestly equipped with playground facilities. A half-dozen youngsters in the age group of four to eight were romping around, chaperoned by two well-dressed young ladies, one of whom looked like the President's sister. A slender young man joined the group and stooped to chat with the children. When he straightened up and turned around, I could see it was the President. He spent perhaps five minutes with the children and then disappeared into the executive offices.

I looked around the Cabinet room. It was simply but distinctively furnished. The room was dominated by a long table that narrowed at each end in order to facilitate easy discourse. Each of the black leather chairs had a small metallic disk identifying various Cabinet posts. The seats immediately adjoining the President's chair were assigned to the Secretary of State and the Secretary of the Treasury. The President's chair was not a matching one, and was obviously new. It operated on a swivel and had a reclining mechanism.

On the wall opposite the garden windows were two average-sized bookcases containing perhaps three hundred volumes, most of them sets of presidential papers.

The President entered the Cabinet room. He was superbly tanned and radiated good health and spirits. I thought back to the time, twenty-six years earlier, when I saw President Roosevelt at the White House and I recalled how struck I was with F.D.R.'s buoyant physical appearance. I remembered thinking that I had never seen a man who had seemed as fully alive as F.D.R. Now, looking at J.F.K., I was reminded of that earlier experience.

The President took me into his office and said he had been fully briefed by Ralph Dungan. One of the principal hopes for world peace, he said, was that the leaders of the Soviet Union would continue their break away from Stalinist habits, suspicions, goals. There was no alternative to peace among the great nations, especially between the Soviet Union and the United States. He said he believed in the need for creating genuinely amicable relations with the Soviet Union. He hoped we had been through the worst of it with Cuba.

The President said the Russians had miscalculated badly in Cuba. They had assumed we intended to invade.

"We never had any intention of invading Cuba," the President added. "Certainly there were those who advocated an invasion but I decided against it for one simple reason: it would have killed too many Cubans. This was why we didn't commit our forces in the Bay of Pigs episode. Anyway, the Russians made a serious error in their estimate of our intentions."

In any case, he said, the important thing now was to get on with the business of reducing tensions. One immediate positive measure that might be taken was an agreement to outlaw nuclear testing. But, he added, Russian leaders seemed overly suspicious and held back in agreeing even to the minimal inspection that would have to be part of any such comprehensive test ban.

The President got out of his rocking chair and walked over to the window. He was reflective. After a moment he turned and said, "You'll probably be talking with Mr. Khrushchev about improving the religious situation inside the Soviet Union, and I don't know if the matter of American-Soviet relationships will come up. But if it does, he will probably say something about his desire to reduce tensions, but will make it appear there's no reciprocal interest by the United States. It is important that he be corrected on this score. I'm not sure Khrushchev knows this, but I don't think there's any man in American politics who's more eager than I am to put Cold

War animosities behind us and get down to the hard business of building friendly relations."

Before I left I was given a letter which asked me to convey the President's Christmas greetings to Pope John and his good wishes for the Pope's full recovery from his illness, the seriousness of which was not known at that time.

I left for Rome on December 1, 1962. It was a comparatively slow flight, with stops in the Azores and Spain. I had a chance to think quietly and consecutively. Could anything be more improbable than attempting the job of messenger between the Vatican and the Kremlin? One point that troubled me had to do with Father Morlion's belief that the time might now be propitious for seeking an amelioration of the religious situation inside the Soviet Union. Did this mean amelioration for Catholics only, or was the attempt to be in behalf of all religions?

When I arrived in Rome late in the afternoon Father Morlion, rotund and beaming, was at the gate to the terminal. He was accompanied by Monsignor Don Carlo Ferrero, executive vice president of Pro Deo University and Morlion's associate and confidant. On the drive to the hotel we discussed the plans for my meetings with Vatican officials, beginning that evening with a visit to the home of Monsignor Igino Cardinale, Chief of Protocol in the Vatican Department of State. Then, on the next day, there would be separate meetings with Archbishop Angelo Dell'Acqua, Deputy Secretary of State, and Cardinal Augustin Bea, president of the secretariat in charge of relations of the Ecumenical Council with non-Catholics.

After dinner at the hotel that evening we called on Monsignor Igino Cardinale. He spoke English with a distinct American accent—the result, I learned, of an upbringing in

Brooklyn, New York. He had come to Rome in 1938, where he worked as a chaplain in various parishes from 1941 to 1946 when he was appointed secretary to the Apostolic Delegation to Egypt, Palestine, Transjordan, Arabia, and Cyprus. He was appointed chief of protocol under Pope John in 1961. He had also written with distinction in the field of papal policy on world affairs, his main work, *Le Saint-Siege et la Diplomatie*, being the only treatise of the kind available in modern times on papal diplomacy.

I showed Monsignor the letter conveying President Kennedy's Christmas greetings to the Pope. We plunged into the matter of the mission to Moscow. He said the time was most auspicious for such an undertaking and that every effort should be made to take advantage of whatever constructive new openings might exist in the Soviet Union. He believed the defeat of Khrushchev's coexistence policies could have serious implications. The changes in the Soviet Union in recent years away from the old authoritarianism should be encouraged in every way. He thought it might be useful to seek some level of representation by the Church inside the Soviet Union—on the assumption, of course, that there would be a genuine improvement in the religious situation inside the Soviet Union.

Monsignor Cardinale stressed the need for total secrecy of the mission. He explained that if the story broke, it would probably be necessary for all sides to repudiate it.

In response to my question about the Pope's health, the Monsignor confided that the Holy Father's illness was not a temporary indisposition, as had been reported in some newspapers, but a painful and malignant disease. The Pope showed physical evidence of the suffering but astounded the men close to him with his determination to carry on the main part of his work.

Before I left, Monsignor Cardinale said that he would pick me up at my hotel in the morning and take me to the Vatican for a meeting with Archbishop Dell'Acqua.

The following morning Monsignor Cardinale came for breakfast. Then we drove into the mammoth cobblestoned courtyard of St. Peter's. The car turned into a narrow roadway flanked by heavy stone walls and barely wide enough for a single vehicle. The driveway opened out again into the interior court of the Vatican. We entered one of the many doorways and took a small elevator to the third floor. The Monsignor's office was in a section occupied by the Department of State of the Vatican. The approach to the office was through a wide, high-ceilinged outdoor colonnade facing the courtyard. On the arched ceiling were paintings by some of the world's greatest artists. The paintings were weather-beaten and faded, having been exposed to the elements for decades. When I mentioned this to Monsignor Cardinale, he groaned and spoke of the difficulty in protecting the paintings. He said there was no way even of approximating the worth of the paintings or the loss represented by weather damage.

The Monsignor said the Pope was about to attend a brief canonization ceremony before the Ecumenical Council. He said he would be glad to take me to the ceremony if I wished. This was something I didn't want to miss, and I said so.

In that case, the Monsignor said, we could recess our discussion and proceed immediately to the cathedral, where the ceremony was about to begin. In fact, we would have to hurry. He would have to take me through the back passages.

He led me swiftly through secret, dimly lit corridors—so narrow we had to scurry along in single file, the Monsignor's cape flying up behind him. The clatter of our heels set up weird reverberations and I had the feeling of being part of an old horror movie. We darted through several private chapels, finally coming to a small door that opened directly into an alcove of the great cathedral, close to the section where the Monsignor wanted me to be seated.

Upward of ten thousand people were massed under the dome. Opposite me, in full splendor, were the Cardinals. Then, behind them, and stretching far back, row after row, were the purple-caped bishops. Nearby was a small tier of Dominican monks. I heard a cry go up from the back of the cathedral; it rolled forward, multiplying in volume. The Pope had entered, borne aloft in the Papal chair. The acclaim was freshened by new waves of tumultuous greeting as he was carried forward. When the procession passed directly in front of me, I could see the Pope's face clearly. He smiled as he waved the Papal benediction to worshipers on both sides of him.

Then came the procession, full of splendor and color and led by the Cardinals in full Church robes. The brief ceremony that followed was almost an anticlimax. I was then escorted back to Monsignor Cardinale's office.

The Monsignor said that his superior, Archbishop Dell'Acqua, was waiting to see me. The Archbishop, a robust man in his mid-fifties, began the meeting by saying he understood I had come with a Christmas greeting for the Holy Father from President Kennedy. He added that, in view of the Pope's condition, it would not be possible to see him at this time, but a meeting could probably be arranged on my return from Moscow. He said it was of the utmost importance for the

Church to take into account the many new changes inside the Soviet Union under Khrushchev. If these changes meant that the chances for averting a nuclear war were improved, then it was natural and right that these changes be recognized and welcomed. And if, furthermore, the changes meant there was any prospect for an improvement of religious conditions inside the Soviet Union, that chance couldn't be ignored. He hoped I might be able to make known to Premier Khrushchev the great value placed by the Holy Father on world peace. Also, the Pope was mindful of Premier Khrushchev's statesmanlike action in withdrawing the missiles from Cuba.

"As the Pope said in his message during the Cuban crisis, he will go out of his way to praise any man in government who is able and willing to help spare mankind the holocaust of war. When you see Khrushchev, you must be sure to mention this. It is important to know, too, whether the Soviet Union would welcome further intervention by the Holy Father in matters affecting the peace."

Then the Archbishop discussed the significance of the ideological division inside the Communist world. If the faction that believed in the inevitability of war were to become dominant, this could have ominous implications. The Vatican followed these developments carefully.

The next morning I went to the office of Cardinal Bea. His assistant, Father Schmidt, acted as interpreter. I could see at once that the Cardinal's reputation for being the kindliest of men was well earned. He was slightly stooped, although his intellectual vigor quickly belied his eighty-one years. Like Archbishop Dell'Acqua, he believed that the smallest possibility for bettering the conditions of the Soviet people should be explored. And if the leadership of the Soviet Union genuinely wanted to improve its relationships with the West, this could lead to substantial benefits, both for the cause of peace and the situation inside the Soviet Union itself. In any event, he felt the matter was worth exploring.

The central question, of course, was whether such explorations would be welcomed on the other end. Was there anything specific I might ask for in Moscow that would indicate a positive response? For many years, he said, members of the religious community had been imprisoned inside the Soviet Union. It would be a most favorable augury if at least one of them could be released.

Was there any particular person he had in mind, I asked.

"Yes," he said, "Archbishop Josyf Slipyi of the Ukraine, who has been imprisoned for eighteen years. He is a very fine man. The Holy Father is concerned about him. He is now seventy. There may be only a few more years left to him. The Holy Father would like the Archbishop to live out those few years in peace at some seminary, where he would be among his own. There is no intention to exploit the Archbishop's release for propaganda purposes."

"Is there anything else we might ask for?"

The Cardinal agreed with Monsignor Cardinale and Archbishop Dell'Acqua that this might be a good time to press for religious improvemnet within the Soviet Union. It was difficult to obtain Bibles; religious education was proscribed; seminaries were being closed. Perhaps these matters might be explored.

"Premier Khrushchev probably thinks we want to restore the Church to what it was in pre-revolutionary Russia," the Cardinal said. "Not true. There were many abuses by the

Church at that time. In many respects, it was a terrible situation. This is not our idea of the proper role of the Church."

It seemed to me that this was a good time to express my apprehensions. And so I asked whether the representations were to be made in behalf of Catholics alone or of all religious beliefs. There had been profoundly disquieting reports of anti-Semitism in the Soviet Union. Wouldn't it be strange to discuss the conditions of religious worship in the Soviet Union without referring to this most striking example of discrimination?

Cardinal Bea was emphatic in saying he hoped I would express Pope John's deep interest in this condition of all religious believers in the Soviet Union. He also urged me to make known to Mr. Khrushchev the Holy Father's profound concern over the conditions of Jews in the Soviet Union.

Would it also be agreeable to take up with Mr. Khrushchev the right to publish not only the New Testament but the Old Testament, Koran, and other holy books?

"That goes without saying," Cardinal Bea said.

Early the next morning I left Rome, changing planes in Zurich, and arriving in Moscow shortly after noon. Waiting to greet me at the airport were two of my Dartmouth conference colleagues, Grigory Shumeiko and Evgeny Fedorov.

CHAPTER
THREE

*T*wo FULL DAYS IN Moscow before the meeting
with the Chairman gave me a chance to discuss matters with
Shumeiko, Fedorov, and Yuri Zhukov. I was especially eager
to learn as much as I could about the present position of the
government on such questions as the test-ban issue and the
ideological dispute with China. I saw Zhukov at the offices of
Pravda, of which he was then associate editor. Zhukov, who
had a long continental background and who spoke French
fluently, was a student of artistic and cultural developments
in the West.

Pravda's building in Moscow was not so heavy or squat
as most of the traditional Russian architecture. Its lines were
comparatively modern and seemed well suited to a communications enterprise. Zhukov's office was large but unpretentious.
Its cluttered, random quality gave it a ready kinship with
editors' offices all over the world.

Zhukov lost no time in bringing me up to date since we
had last spoken at Andover. He said the Cuban denouement had
created a precariously balanced situation. Substantial hope
had developed after Cuba for a series of agreements with the
United States that might justify Khrushchev's Cuban withdrawal. The Chinese had seized upon the withdrawal as proof
that Nikita Khrushchev was unable to stand up to Western

imperialists, and had, in fact, capitulated. Khrushchev, the Chinese contended in their propaganda, had not only sold out to the West but was actually leading the Russian people away from socialism with his bourgeois revisionism. I asked Mr. Zhukov how much of an impact this was making inside the Communist world.

In most places, very little, he said. Even so, it became necessary for the Chairman to vindicate his basic policy of coexistence and to demonstrate that the Cuban situation, far from representing capitulation, could lead to agreements with the United States.

In surveying the possibilities of such agreement, he said, three areas came to mind in which both countries were not too far apart. One area was nuclear testing. The Soviet Union would withdraw its objections to on-site inspection so long as it had reasonable assurance that inspection would not be used for espionage purposes.

The second area of possible agreement concerned outer space and the need to keep it free of weaponry. He anticipated no difficulty in this matter.

The third area was the Berlin situation. The Soviet Union was opposed to the presence of American troops in Berlin. He recognized it was unrealistic to suppose that the United States would withdraw its forces so long as the general situation in Berlin was unstable. But it was now felt that a formula could be drawn up to satisfy both the Russian and American positions. The way to do it would be by establishing a United Nations presence in Berlin. American troops might then remain under a United Nations flag, but at least the presence of the United Nations could help lift Berlin out of the context of the cold war.

In any event, it was the hope of the Soviet leaders, he said, that agreement in these three areas could be speedily achieved.

After leaving the offices of *Pravda* I met with Oleg Bykov, whom I had known since the first American-Russian conference at Dartmouth in 1960, where he had served as aide to Alexander Korneitchuk, chairman of the Soviet delegation. He said that Korneitchuk was then in a hospital near Moscow and was convalescing from a bronchial ailment that had kept him from attending the Andover conference. Bykov asked if I would like to visit Korneitchuk.

Korneitchuk's private hospital room was bright, commodious, well furnished. Korneitchuk was waiting at the door of his private room, looking dapper in an elegant bathrobe. He joked about his health, saying that his incarceration in the hospital was a plot by rival playwrights to keep him out of competition. He had foiled them by writing two plays while on his back. He had some chest pains, he said, but he was not uncomfortable. He had been promised a complete cure; he had a little vodka in his closet to prove it. He poured drinks, then summoned the nurse to set a table of fruit, tea, and cake.

He confirmed everything I had heard about Khrushchev's need to produce effective agreements with the United States in the wake of the Cuban episode. Khrushchev's supporters felt that the decision to withdraw from Cuba was an act of statesmanship and high responsibility, and could represent a vital turning point in the cold war. Others, however, reserved judgment, saying it would be necessary to come up with specific agreements before such an optimistic interpretation could be sustained.

I knew Korneitchuk to be a confidant of the Chairman. What would happen, I asked him, if Khrushchev failed to obtain the agreements he needed to justify his Cuban policy?

It was important to understand, Korneitchuk said, that Khrushchev was irretrievably committed to two objectives. One was to upgrade the living conditions and spirits of the Russian people. The second was to eliminate the dangerous feelings of suspicion and hositility toward the West, implanted under Stalin. Specifically, this meant ending the arms race.

The two objectives were related, he said, because only by lifting the burden of arms manufacture would it be possible for the Soviet economy to be freed for economic development and improved living conditions. Coexistence, therefore, was not just a matter of foreign policy; it was directed to the most important need of the domestic economy itself.

Khrushchev would not easily be deflected from these commitments, he said. Even if he didn't get the agreements he thought would be acceptable to the United States, he would persist with his policy of coexistence as long as he had enough support inside the country.

I didn't want to tire my host, but there was one small point I thought I might bring up. I would consider it a great favor, I said, if I could bring my own interpreter to the meeting with Mr. Khrushchev. Not that I lacked confidence in the person Mr. Khrushchev might designate. It was just that I thought it might be more satisfactory if I could talk to the Chairman through an interpreter who was accustomed to my idiosyncracies.

Did I have any particular intrepreter in mind? he asked.

Yes, I replied, pointing to Oleg Bykov, who was completely familiar with the idiomatic use of the American language. (Russian interpreters tend to be divided into two groups: those whose English was learned in England or under

English-taught instructors, and those who speak with a distinct American style of speech. Bykov belonged to the latter.)

Korneitchuk said he would see what could be done and would inform me at the hotel.

The next morning Oleg Bykov telephoned to say that the Supreme Soviet—equivalent of Parliament or Congress—would be in session that day. Premier Khrushchev would be addressing the delegates at 2:00 P.M. That would be Mr. Khrushchev's first public accounting for his decision to withdraw from Cuba. He would probably reply to the criticisms of the Chinese leaders. Chairman Khrushchev thought I might be interested in going. Would I accept his invitation?

Of course.

That afternoon Oleg escorted me to one of the largest buildings in the Kremlin complex. We ascended a long, red, deep-carpeted stairway to a larger foyer adjoining a rectangular, simply designed auditorium where the representatives from the various districts were taking their seats for the afternoon session. Oleg took me directly to the Kazahkatan section toward the front of the hall.

Shortly after we were seated two women walked up the aisle and took their places three rows in front of us. Oleg identified them as Mrs. Khrushchev and the wife of Marshall Tito. A minute later Nikita Khrushchev came to the platform. An ovation ensued. After returning the applause in Russian style, Mr. Khrushchev began to talk over the clamor and the hall soon became quiet, filled only by the Chairman's staccato delivery. The first part of his talk was directed to economic conditions inside the Soviet Union. Then he entered into a review of foreign affairs, and proceeded to the matter of Cuba.

He said the decision to equip Cuba with modern weapons

was dictated by American plans for invading Cuba. Then, as the crisis developed, he said, it became clear that the Cuban situation was getting out of hand and that a terrible culmination was building up. Both he and President Kennedy had the joint responsibility to prevent the Cuban crisis from leading to nuclear war. He said the President had assured him that the United States had no intention of invading Cuba. Therefore he felt it was right to remove the missiles. But the Chinese were attempting to make it appear that a sober, responsible decision to avert nuclear war was a repudiation of Marxist-Leninist ideology. One could only be astounded, he said, by such tortured logic. It was like being in a fight and having someone on the sidelines goad you on to your own destruction. He didn't regard the responsibility given to him by the Party as a license to help destroy the human race. Marx and Lenin were no freebooters or military adventurers. It would not be possible to pursue the triumph of socialism on the radioactive ruins of civilization. As for the derisive comment of the Chinese that the United States was bluffing and was only a "paper tiger," he commented that the tiger had nuclear teeth. He insisted that the Soviet Union would not let the Cuban people down and intended to stand by its commitment. Meanwhile, he said he would do everything he could to seek a peaceful resolution of the Cuban situation and all the other critical situations in the world. The remainder of the Chairman's talk was largely devoted to ideological matters.

On the way back to the hotel I asked Oleg what he thought of the talk.

"Well, you were able to see for yourself," he said. "The old man really means it. And the people know he means it. I think the Chinese know he means it, too.

"Incidentally," he added, "I've been assigned as your interpreter. Korneitchuk made the suggestion and got an immediate okay."

For five hours that evening I rehearsed with Oleg for the meeting the next morning with the Chairman. I wanted to be sure not only that everything I would say to Khrushchev was completely intelligible to Oleg for interpreting purposes but that the emphasis would be accurate. I encouraged Oleg to ask questions about shadings in meaning—something that would be awkward during the session with the Chairman. It developed there were at least three dozen terms where precise equivalents were lacking in the Russian language, and it became necessary to develop context to convey meaning. Exhausted but hopeful, we ended our session at 1:00 A.M., after agreeing to meet at breakfast for a final go-round.

The next morning we tried to anticipate some of the questions Mr. Khrushchev might ask. At 10:15 A.M. we left the hotel for the 11:00 A.M. appointment at the Kremlin.

Nikita Khrushchev had his offices in a building of prerevolutionary vintage on a Kremlin side street not open to the general public. The main doorway was so unpretentious that I thought for a moment we were using the back entrance. The small foyer was more suggestive of a lobby in a modest apartment house than the reception hall of the headquarters of a major government.

Bykov and I were ushered into a small anteroom. We were hardly seated when Mr. Khrushchev came to the door, greeted us, and escorted us to his adjoining office, the most conspicuous feature of which was a long, narrow conference table that could accommodate twenty persons or more. The table was pressed against Mr. Khrushchev's desk and made a T design.

I introduced Oleg Bykov.

"So this is the famous interpreter you bring with you," Mr. Khrushchev said. "Tell me, Mr. Bykov, have you ever sat in the Prime Minister's chair before?"

Oleg mumbled what I took to be the equivalent of, "No, sir."

"Very well," said the Chairman. "You'll now see how it feels. I will sit opposite Mr. Cousins at the conference table and you will be at the Prime Minister's desk in the center. If the chair's too uncomfortable, I will requisition a better one."

Oleg assured him he would be most comfortable. We took our places.

"Now," said the Chairman, addressing me, "we will have man-to-man talk. Please tell me about your family. In Russia we like to hear about families before we talk about business."

I spoke about my wife and four daughters. The Chairman asked if I had brought them with me. When I said I had not he looked at me severely and said: "For shame."

I explained that the girls were at school and he dismissed what I said with a wave of the hand.

"School? Nonsense! They don't teach anything in the schools as important as they could learn traveling with Papa. Weren't they even curious about your trip?"

I said that the youngest, then twelve, asked me before leaving whether I was terrified about being alone with the mighty head of all the Communists. Then, without even waiting for my answer, she said, "Daddy, when you see him, just imagine he's an old uncle and you won't be scared any more."

"My grandson gives me good advice, too," Mr. Khrushchev said. "In fact, sometimes I make decisions that members of the Party say they don't fully understand. When they ask me how I happened to decide, I tell them this is what my grandson told me to do. They think I'm joking. They don't know how wise my grandson is."

I noted that Mr. Khrushchev had had a busy time with the Supreme Soviet, and thanked him for the opportunity to sit in on the session of the Supreme Soviet at which he spoke. Was he exhausted?

"It's really not too bad—but I am a little tired now that it's over," he said.

Then he asked if I had heard Tito.

I said no. Tito had spoken at the morning session.

"The Tito matter gave me a few more gray hairs," he said. "As you know, our relations with Yugoslavia are rather delicate. I don't want Marshal Tito to think that we are too big to have equality with him in our relations. We have our differences, and I've tried very hard to persuade him I respect his position on these differences and that we're not trying to gloss over them.

"So yesterday, in introducing Marshal Tito to the Supreme Soviet, I made it clear that, even though our two countries disagreed on certain matters, we were not allowing these differences to stand in the way of our friendship.

"Marshal Tito acknowledged the introduction very cordially and spoke along the same lines as I did. He received a very warm response at the end of his talk. Then there was a recess for a few minutes. It had gotten pretty stuffy and I needed air. I went out for a walk.

"One thing I like to do while I walk is listen to music. I

have a small transistor radio I keep in my pocket. And so yesterday, as I took my walk, I held the tiny radio to my ear, listened to the music, and tried to clear my mind. The music was interrupted for a news bulletin. The announcer reported on the morning session of the Supreme Soviet just completed. He said Marshal Tito and I had spoken to the Supreme Soviet and that we had announced that all differences between our two countries had been fully resolved. I could hardly believe my ears. We had done nothing of the sort. What really troubled me, of course, was that if this news report came to the attention of Tito, he would think I was playing a double game —saying in his presence that I recognized the fact of the important differences between us but was telling a different story to the press, glossing over our differences and making it appear they no longer existed. How could that idiot of an announcer have said such a stupid thing? What happened, of course, is that some journalists just don't know how to handle good news.

"Now, what should I have done? Should I have rushed back to Tito and apologized? It was possible he knew nothing about it. Why confuse him? In fact, he might be even more upset by the apology than by the news bulletin. But if I did nothing, wouldn't I be taking a chance that somebody in his entourage might have heard the broadcast and elaborated on it in the retelling, with the result that all our efforts to establish good relations with Tito would be jeopardized? Besides, wouldn't he assume that the news story was official if I didn't repudiate it?

"Finally, I decided to wait twenty-four hours. I figured that if he or his aides had heard the broadcast, he would be certain to protest. In that case, I could say, 'Yes, I know it's a stupid, dreadful thing. Please pay no attention to it. I'm investigating to find out how it happened and I intend to give you a report.' But if, at the end of twenty-four hours, he says nothing to me, the chances are he knows nothing about it and there's no problem. So, I've still got an hour to go. The things a man gets into when he gets into politics. . . ."

He shook his head.

"But you didn't come here to hear me complain about my problems, especially after having listened to me for two hours yesterday. That's a long time for a capitalist like you to listen to an old Communist like me. I hope I didn't shake you in any of your beliefs."

I said I had listened with the keenest interest to everything he had to say, and was glad to reassure him he had not deprived me of my philosophical underpinnings. I added that I was pleased to hear him tell the Supreme Soviet that there had been an economic upturn.

The Chairman said his country had been making progress. In industry they had surpassed their quotas. That didn't satisfy him. The quotas should have been set much higher. But at least the increase in industrial production gave promise of much larger gains ahead. In agriculture they hadn't done nearly so well. They were ahead of the previous year but still far behind where he thought they ought to be. One thing they had done might help, he said. They had just divided the Communist Party into two major sections—one industrial, one agricultural. In this way he thought they ought to be able to sharpen the lines of responsibility.

But they were still plagued by bureaucracy, he continued. The bureaucracy made for inefficiency. If something went wrong, it was always someone else who was responsible.

The bureaucracy and the incompetence didn't happen overnight; they couldn't be eliminated overnight. They were built
into the way things were done during the long years under
Stalin. The Chairman said people had a habit of finding easy
excuses for doing the wrong things. He would be told this was
the way matters had always been and that it was the only way
people knew how to do them. He realized, therefore, that
there would have to be something approaching a psycholog-
ical upheaval before people would be ready to face up to the
need to change the way they were doing things.

This meant de-Stalinization. The bureaucracy had grown
up under Stalin. Only by changing the attitudes toward Stalin
could they change everything else that had to be changed, he
said. But this was a real problem. Stalin had been worshiped
by the Soviet people. Millions of people went off to war and
died with the name Stalin on their lips. They had no idea how
irresponsible and irrational he was. Did anyone have the right
to disillusion them? Wouldn't there be a profound emotional
shock if they were told that the man they had venerated for
so long didn't really know what he was doing?

"I wrestled with the problem," he said, "then finally decided I had to tell the people the truth. At least twice I made
long statements on the subject, telling the full story. You would
suppose that by now people would know. Not so. Every day
I meet otherwise intelligent people who still think Stalin was
sane.

"There was a very important difference between Lenin
and Stalin. Lenin forgave his enemies; Stalin killed his friends."

All the time he spoke, his hands were folded quietly on the
table. I had expected him to be volatile and free swinging in
manner, judging by the newsreel pictures showing him en-

gaged in extravagant gesturing and posturing. Yet his private demeanor couldn't have been more restrained or polite. He spoke in even, subdued tones. Even his clothes added to the impression of restraint. He wore a dark blue suit, white silk shirt, solid gray tie held in place by a small jeweled stickpin. His shirt had French cuffs with large gold links. The break in the cuffs revealed long-sleeved winter underwear. The elegance of his attire compared to my own made me feel somewhat awkward. I looked down at my unmoored tie and my plain cuffs, then tucked the tie inside my jacket.

The Chairman said he was happy to accept the suggestion of several of the Soviet delegates who had been to the Andover conference in the United States that I be invited to come to speak to him. He said he had seen various materials prepared by Father Morlion that had been sent to him in advance of our meeting. Then he said he understood I had just come from Rome.

"What can you tell me about the Pope?" he asked. "Is he very ill? He made a big contribution to world peace during that terrible time of the Cuban crisis."

I said that, despite his severe illness, Pope John was determined to use his remaining energies in the cause of peace. I emphasized, of course, that I was speaking not as the official emissary of the Pope or of the Vatican in general. I had come in a private capacity but I had seen Vatican officials and was in a position to convey my impressions.

"I understand completely," the Chairman said. "No one need be committed. About the Pope: he must be a most unusual man. I am not religious but I can tell you I have a great liking for Pope John. I think we could really understand each other. We both come from peasant families; we both have lived close to the land; we both enjoy a good laugh. There's something very moving to me about a man like him struggling despite his illness to accomplish such an important goal before he dies. His goal, as you say, is peace. It is the most important goal in the world. If we don't have peace and the nuclear bombs start to fall, what difference will it make whether we are Communists or Catholics or capitalists or Chinese or Russians or Americans? Who could tell us apart? Who will be left to tell us apart?"

His eyes were in a vacant stare.

"During that week of the Cuban crisis the Pope's appeal was a real ray of light. I was grateful for it. Believe me, that was a dangerous time. I hope no one will have to live through it again. Well, I think you know how I feel about it. You heard me speak about it yesterday."

Judging from the Chairman's remarks on that occasion, I said I had the impression he felt there were misunderstandings in his country about the Cuban crisis that called for correction.

"Not so much inside this country, but the Chinese have done everything they can to misrepresent what happened. You get echoes of their propaganda here and there. The Albanians, for example.

"The Albanians remind me of something that happened when I was young and used to work in the mines," he continued. "When the miners came up for lunch, they would amuse themselves by calling over small children and offering them kopeks if they would memorize some words and then go home and recite the words to their parents. When the children agreed and held out their hands, the miners gave them kopeks and proceeded to instruct the children in the foulest words known to the Russian language. And afterward when the kids left, the miners would howl with laughter when they imagined these tots teaching their parents such choice vocabulary.

"It was stupid, of course, but this was the sort of thing that amused the miners. Well, the point of my story is, somebody has been teaching the Albanians dirty words and giving them kopeks."

Among the "dirty words" used by both the Chinese and the Albanians to describe Khrushchev's Cuban policy were "cowardice" and "inability to stand up to the American paper tiger." Mr. Khrushchev himself, in his talk to the Supreme Soviet, referred to the allegation that he had been badly frightened, and then said that any man who could stare at the reality of nuclear war without sober thoughts was an irresponsible fool.

"How did it feel to have your fingers so close to the nuclear trigger?" I asked.

"The Chinese say I was scared. Of course I was scared. It would have been insane not to have been scared. I was frightened about what could happen to my country—or your country and all the other countries that would be devastated by a nuclear war. If being frightened meant that I helped avert such insanity then I'm glad I was frightened. One of the problems in the world today is that not enough people are sufficiently frightened by the danger of nuclear war.

"Anyway, most people are smart enough to understand that it is ridiculous to talk in terms of another war. Pope John understands this. I would like to express my appreciation to him for what he did during the crisis of the Cuban week. Do you have any suggestions?"

I said I was certain that the adoption of policies that would make for genuine peace on earth was the finest reward that might be given the Pope. Naturally, the Pope was profoundly interested in the possibility of improvement in the conditions of religious worship under the Soviet Union. If there were new developments indicating such improvement, I was certain he would be gratified to hear about them.

The Chairman said nothing for a moment, then leaned forward in his chair. "Your government in the United States has been separated from the Church," he said, "and you have no idea what the situation was here under the Czar. I can tell you that all of us who lived under it will never forget what it was like. The Church became the means for perpetuating political tyranny and cruelty."

Again I stressed I was speaking as an individual; I felt justified, however, in saying there was no desire to restore the Church to its Czarist status. In fact, there was a keen awareness of abuses that had been carried out at that time. What was sought now was an amelioration of the conditions of religious worship inside the Soviet Union, with full realization that this would be done within the framework of the existing political authority.

He replied that many of the Soviet leaders had strong religious backgrounds. Some of them had even studied in seminaries. They had to struggle against the social injustices and political tyranny of the Czars. He said they saw the Church as a full partner of the regime. "The priests were the gendarmes of the Czar, and they had to be uprooted along with everything else that belonged to the Czar."

Once again, I emphasized, there was no idea of reverting back to Czarist traditions. But religious freedom was guaranteed under the Soviet constitution. Therefore there was nothing inconsistent between what was being sought and Soviet law. For example, there was a need for increased availability of holy literature, release from prison of religious personalities, greater freedom with respect to religious education, removal of difficulties in baptism, eradication of anti-Semitic practices, etc.

The Chairman said he would like to go at these specifics one at a time. He said he wasn't too familiar with the precise situation pertaining to publication of Bibles or other religious literature, but would be glad to look into the matter and review it. Then he asked what was meant by the release of religious personalities from prison.

I said that over the years many attempts had been made to obtain the release from prison of Archbishop Slipyi of the Ukraine. Pope John was hopeful that something could now be done. He was not addressing himself to the reasons for the internment; these reasons went back many years and there was no point in rearguing the case. After eighteen years, however, it was not unreasonable to ask whether the Archbishop should not be given an opportunity to live his few remaining years as a free man.

I thought I detected a stiffening in the Chairman's manner.

"You know," he said, "I'm rather familiar with the Slipyi case. I'm from the Ukraine. The entire matter is still fresh in my mind."

Then, for almost twenty minutes, the Chairman proceeded to describe the religious situation in the Ukraine before 1947. He spoke of the competition between the Ukraine Rite Catholic Church, to which Archbishop Slipyi belonged, and the Russian Orthodox Church. He spoke of the struggle for power inside both groups. He traced the leadership in the

Ukrainian Rite Catholic Church under Archbishop Sheptyt-sky. He told of the meeting in 1946 which resulted in deep divisions within the Ukrainian Rite Catholic Church. He said that Archbishop Sheptytsky died in 1944 under circumstances that indicated "his departure from this earth may have been somewhat accelerated."

In any event, Slipyi had succeeded Archbishop Sheptyt-sky as Archbishop in 1944. The reason for his imprisonment, the Chairman said, had to do with collaboration with the Germans during the war. He added that those who defended Slipyi claimed that "collaboration" was too strong a word and that he had been responsible for saving many lives because of his position.

Once again I said that I hadn't come to argue the original case, but it was now eighteen years since the Archbishop was first imprisoned.

Again the Chairman shook his head. "It is not a good idea," he said. "I would like to have improved relations with the Vatican but this is not the way to do it. In fact, it would be the worst thing we could do. It would make a terrible stink."

When Oleg used the word "stink," I was certain something had gone wrong in translation. I asked for clarification. Oleg said the term "stink" was perhaps a little strong in translation and that what the Chairman meant was that the release of the Archbishop would produce exactly the opposite effect from the one hoped for.

In what respect? I asked.

The Chairman said if the Archbishop became free there would be large headlines proclaiming, "Bishop Reveals Red Torture." Of course, he said, such stories would be false but newsmen were certain to exploit the Archbishop's release in those terms. The net effect would be to worsen relations with the Vatican.

"I think I can assure you," I said, "that Pope John is not seeking the Archbishop's release for purposes of making propaganda against you. The Church is not lacking in materials for this purpose. All the Pope wants is to give Archbishop Slipyi a chance to live out his life in some distant seminary. The Pope is acting in good faith in seeking the Archbishop's release. Incidentally, you probably are aware that Pope John has made no denunciations against you or your government. He recognizes that important changes have been made from Stalin's time and he feels there is hope that this trend can be continued and expanded."

"Let me think about this," said the Chairman. "It is not an easy question. Anyway, as I say, I welcome the opportunity to have good relations with the Catholic Church. This doesn't mean that I'm going to become a Catholic any more than the Pope is going to become a Communist. I'm not going to try to convert him and I know he's not going to be able to convert me, although"—here he grinned—"stranger things have happened. Anyway, I have no objections to the Church so long as it keeps out of politics. In fact, I believe the government should help to protect the Church so long as it stays within its religious purposes. We had an important meeting recently and I took pains to invite the head of the Russian Orthodox Church."

I said I was certain that Vatican officials would be glad to learn of his opinions in these respects and that I would explore the possibilities of improved communications. I did, however, anticipate one problem in this respect. I then referred to the apprehension among some Vatican officials that such improved contacts might be exploited for political reasons.

"Not so," he said. "You can give them reassurance on this point."

As an example of what I had in mind, I referred to a prominent item appearing in the Soviet press shortly after the Cuban crisis. It praised the Pope's call for peace but then proceeded to interpret his action as proof that the Pope was turning against the West and against the United States in particular. Such an interpretation, I said, was inaccurate and harmful. If further activities by the Pope on behalf of peace were going to be exploited for propaganda purposes favorable to the Soviet Union, then obviously the Vatican would have to dissociate itself from such an interpretation.

Mr. Khrushchev said that he was sorry to hear about this news item. He had not seen it. He said it didn't represent his view or the view of the government. He asked, if the matter came up in my conversations with Vatican officials, that his regrets be made known. In the Soviet Union, as elsewhere, he said, it was difficult to keep bumbleheads out of important news jobs—as his own experience with the Tito episode had just demonstrated.

At this point I thought it important to reiterate that the Pope was not asking for improvement in the religious situations of Catholics only. He was speaking in behalf of all religions. This led me to the next major point, anti-Semitism.

I mentioned the increasing restrictions in worship, economic repression, and discrimination in government.

For the first time in our talk I could see that the Chairman was somewhat impatient.

"Not this again!" he said. "I wish I knew how these accusations originated. They're plainly false. Even Mrs. Roosevelt would write to me about anti-Semitism in the Soviet Union.

"Let's start with one fact," he went on. "Is there prejudice against minority groups in the Soviet Union? Yes, there is. But we are working on it. It is not a major problem. Despite all our laws against discrimination and prejudice some prejudice and discrimination exist. This should not be difficult for an American to understand. You have laws in the United States. Despite these laws, millions of your dark-skinned citizens don't have equal job opportunities, equal education, equal rights under the law. There are some cases where law takes a long time to become fully effective. But everything begins with the law.

"We have laws in this country against racial and religious discrimination. By and large, we think we do a good job of enforcing these laws. We like to think we have a much better record than the United States in this respect. Your problem of racial discrimination and segregation, not just in the South but in the North, is a very serious one. And so far your law enforcement has been unable to do the job adequately. You have made progress. But, as I say, everything begins with the need for people to know where its government stands on important questions.

"On the matter of anti-Semitism, our government is officially opposed to all religious or racial discrimination or segregation. Does the government itself discriminate? No. One of the most important jobs in government—the job of finance minister—is held by a Jew. Jews are eminent in other fields—in literature, the theater, ballet, science.

"Let me tell you something else. I'm the grandfather of a Jewish boy. My son married a Jewish girl. They had a child.

Then my son went off to war and was killed. The mother and child became part of my family. I brought the child up as my own. You see how preposterous it is to say that I'm anti-Semitic?"

I asked the Chairman whether he would welcome correspondence on the subject of official and unofficial discrimination against Jews. I had seen abundant documentary material I should like to call to his attention.

"Certainly. Any time." He spoke with emphasis.

As to Pope John's interest in furthering conditions for world peace, I said, the Pontiff believed it might be useful to follow up the success of the International Geophysical Year with an International Cooperation Year. Would the Chairman be willing to support such a project if it were proposed through the United Nations? Mr. Khrushchev replied that it sounded like an excellent idea and that he could see no reason why his country would not give every encouragement to the project.

I stood up to leave. I was mindful of the fact that we hadn't even discussed the matters of concern to President Kennedy. But I was also mindful that the Chairman hadn't had his lunch, even though we had been talking for nearly three hours and it was now almost 2:00 P.M.

The Chairman was reading my mind.

"Please sit down," he said. "How is President Kennedy?"

I said he was in excellent health and spirits. Also, that he was extremely eager to develop the kind of relations with the Soviet Union that would help create the conditions for a more peaceful and orderly world.

The Chairman said he would meet the President, or anyone else, more than halfway for that purpose. For example, he had recently seen President Fanfani of Italy. During the course of an amicable conversation he said to Fanfani, "You belong to the NATO alliance; I belong to the Warsaw Pact. But this cannot, and must not, lead to conflict."

The Chairman folded his hands on the green felt table.

"One thing the President and I should do right away," he said, "is to conclude a treaty outlawing testing of nuclear weapons and then start to work on the problem of keeping these weapons from spreading all over the world. It is not true that I am against inspection. I keep seeing newspaper stories in the United States that the Soviet Union is opposed to inspection as part of any test ban. This is not true. If the United States wants reasonable inspection, it may have it. What we do object to is using a nuclear test-ban treaty as a device for opening up our country to all sorts of snooping that has nothing to do with the test ban. We see no reason why it shouldn't be possible for both our countries to agree on the kind of inspection that will satisfy you that we're not cheating and that will satisfy us that you're not spying.

"Apart from the test ban, of course, there is the problem of Germany. I can understand how Americans look at Germany somewhat differently from the way we do, even though you had to fight Germany twice within a short time. We have a much longer history with Germany. We have seen how quickly governments in Germany can change and how easy it is for Germany to become an instrument of mass murder. It is hard for us even to count the number of our people who were killed by Germany in the last war. More than twelve million, at least. We have a saying here: 'Give a German a gun; sooner or later he will point it at Russians.' This is not just my feeling. I don't think there's anything the Russian people feel more strongly about than the question of the rearmament of Germany. You like to think in the United States that we have no public opinion. Don't be too sure about this. On the matter of Germany our people have very strong ideas. I don't think that any government here could survive if it tried to go against it.

"I told this to one of your American governors and he said he was surprised that the Soviet Union, with all its atomic bombs and missiles, would fear Germany. I told your governor that he missed the point. Of course we could crush Germany. We could crush Germany in a few minutes. But what we fear is the ability of an armed Germany to commit the United States by its own actions. We fear the ability of Germany to start a world atomic war. What puzzles me more than anything else is that the Americans don't realize that there's a large group in Germany that is eager to destroy the Soviet Union. How many times do you have to be burned before you respect fire?"

I asked the Chairman whether he didn't recognize that the danger of a rearmed Germany was part of a larger danger; namely, the fact of a lawless world in which each state determines the requirements of its own security, the net effect being world anarchy and a stage for world war. Wasn't a strengthened United Nations the only true source of security —for his nation or any other?

He replied that the Soviet Union didn't believe the United Nations was a truly objective agency for peace. The influence of the United States in the United Nations was disproportionate, he said; therefore, the United Nations tended to become an instrument of American foreign policy rather than an impartial world organization.

I asked him whether it was pertinent to point out that the United States had similar misgivings about the main agency of the United Nations charged with maintaining world security—the Security Council—precisely because of the influence of the Soviet Union in that body through its veto. Rather than get into a debate on the weaknesses of the United Nations, I wondered whether he would agree that what was necessary was a determined effort to make the United Nations adequate, objective, and impartial in those respects having to do with legality and enforcing world peace.

"Do you expect me to agree to a United Nations that can come into our country and tell us what to do?"

"Certainly not. Quite the contrary. The purpose of United Nations should be to protect the essential sovereignty of nations, large and small, by having adequate authority in matters concerned with the common security of all nations."

"We're not against any idea that really makes for genuine peace," he said, "as long as no one tries to take away the gains of our revolution."

I reverted to a point the Chairman had made several minutes earlier when he indicated that the main purpose of the United States in seeking inspection as a condition for a test-ban treaty was to carry on snooping activities.

"In my opinion," I said, "President Kennedy is genuinely seeking an agreement to end testing. He is genuinely interested, too, in improving relations with the Soviet Union."

The Chairman looked up at me, raising his eyebrows as though to say, "Tell me more."

I suggested to the Chairman that many political observers in the United States believed that no one who aspired to the Presidency in either party was more eager than John F. Kennedy to end the animosities of the cold war and to create a basis for constructive relations with the Soviet Union.

The Chairman said that if such were the case, the President "would not find me running second in racing toward that goal."

Again I stood up and started to thank him for his hospitality.

"Before you go, let me give you something for the Pope and President Kennedy," he said. "Just some Christmas greetings." He reached into a drawer and took out some official stationery on the front of which was an embossed drawing of the newest of the Kremlin buildings. Then in his own hand he wrote messages to Pope John and President Kennedy. After he had finished, he handed them unsealed to Oleg Bykov and instructed him to translate the messages.

I found it striking that the message to Pope John used religious terminology not readily associated with the leadership of the Communist party of the Soviet Union. It made a specific reference to the Holy Days in wishing the Holy Father the best of health at Christmastime. The other note was a simple expression of good wishes during the holiday season for the health and well-being of the President and Mrs. Kennedy.

On the way to the door Mr. Khrushchev asked me to convey his Christmas greetings to Father Morlion and to thank him for his part in arranging our meeting; also to tell him that he would be delighted to have him come to visit him in Moscow. "He doesn't have to take off his priest's clothes when he comes." He smiled.

I thanked the Chairman for his cordial hospitality and left.

Early the next morning I was on my way back to Rome.

*M*ONSIGNOR CARDINALE, Father Morlion, and Don Carlo Ferraro stayed up late that night to hear the report on the events in Moscow. Monsignor Cardinale said the Pope's condition was slightly improved and that there probably would be an opportunity for me to report to him directly the next afternoon. Meanwhile, he had arranged for separate meetings with Archbishop Dell'Acqua, Cardinal Bea, and Cardinal Eugenio Tisserant, dean of the College of Cardinals. He was especially eager that I give a full account to Archbishop Dell'Acqua.

The schedule the next morning went off as planned. Most of Cardinal Bea's questions centered on the Chairman's reactions to suggestions for religious amelioration and the request for the release of Archbishop Slipyi. Archbishop Dell'Acqua was interested, of course, not only in the purely religious aspects of the discussion with Chairman Khrushchev but in the possibilities for political and ideological change inside the Soviet Union. On the basis of the report, he said he felt it was more propitious than ever for greater involvement by Pope John in matters concerned with peacemaking among nations. Indeed, he felt that any delay might be costly in the sense that Khrushchev's coexistence policy might be shelved unless he could show results. The alternative to the Khrushchev policy was bound to be carried out by men whose ideological and historical leanings were against the West. He felt especially concerned about the need for agreement on a treaty to outlaw nuclear tests.

Cardinal Eugene Tisserant, dean of cardinals, world-famous not only as a churchman but as a scholar, resembled an Old Testament prophet. His large brown eyes and russet beard dominated a proud head. He listened intently, saying very little. He seemed especially interested in the Chinese developments.

Having made the rounds, I returned to the offices of the Vatican Secretary of State. As we passed through the large doors Monsignor Cardinale told me of an episode that occurred at that spot the evening of October 28, 1958, when the Papal election ceremonies in the Conclave of Cardinals were completed.

Well-wishers in profusion pressed in upon Pope John XXIII. Indeed, the congratulatory urge felt by many of the members of the hierarchy was so strong that aides of the Pope tried to protect him by installing him in one of the offices of the Vatican Secretary of State and placing the Holy Seal across the door.

Breaking the seal is a profound sacrilege. But the enthusiasm of the pursuers, many of whom were cardinals and bishops, persisted, and they swept past the seal as though it never existed. Trying desperately to stem the tide, Cardinal Tisserant cried out to Monsignor Cardinale, then secretary of the Conclave: "Stop them! Do something! Tell them to stop or they will be excommunicated!"

Pope John smiled. "Very well," he said gently. "They will be excommunicated and my first act as Holy Father will be to grant them complete absolution."

Everyone laughed, and the tension was broken. The Pope quickly became engulfed by a human congratulatory tidal wave. It was a strenuous session for the seventy-seven-year-old Pontiff, but he was equal to it.

Monsignor Cardinale then related a number of other anecdotes about Pope John XXIII, all of which pointed up his genius for human relationships. Indeed, the historical Ecumenical Council was a reflection of his desire to bring the Catholic Church into closer contact with the outside world, making it more responsive to the needs of human beings everywhere, whether Catholic or not. He didn't believe that God penalized anyone for not being a Catholic. Religion was a matter of individual conscience. All religions were entitled to respect. Even non-believers who have had audiences with the Holy Father were told that he included them in his prayers.

I learned that Pope John's flair for human relations and the importance he attached to direct contacts with the outside world were highlighted by several specific incidents. Shortly after his election he set out from the Vatican on the first of a series of visits to Italian prisons. Asked by his aides to explain his purpose, he said simply: "It is somewhat more difficult for the prisoners to come to see me."

On another occasion the Pope had left his car and was strolling back to his apartment in the Vatican when a distraught priest came up to him and begged his prayers for the paralyzed wife of a friend. The Pope said he could do better than that: he would go directly to the stricken woman at her home, which he did.

The third incident told me by Monsignor Cardinale concerned Pope John's central purpose. A Canadian dignitary asked the Pope to explain the main objectives of his papacy in general and the Ecumenical Council in particular. Pope John stood up, walked over to the window, opened it, and said, "What do we intend to do? We intend to let in a little fresh air."

It was made clear to me that Pope John had no intention to dictate change; his purpose was to set the stage for it. The Ecumenical Council, already one of the great events in the history of religion, was called for the purpose of having peoples of all Christendom consider what kind of changes were required and how best to meet the problems involved.

Pope John's approach to the Ecumenical Council reflected his conviction that the Church was not the private possession of its hierarchy but a common responsibility shared by all its members. Accordingly, he made an important distinction between dogma in theology and dogmatic attitudes. It was one thing to be strongly rooted in one's religious convictions; it was another thing to be rigid and authoritarian in one's attitude toward human problems and relationships with people of other faiths. A dogmatic attitude or approach toward the honest convictions of one's neighbors was itself a violation of the religious spirit. All men, whatever their beliefs, were important; God did not impose penalties or withhold blessings on people just because they were not Catholics. In this sense, all had access to the Deity.

Monsignor Cardinale told me my appointment with the Pope was scheduled to follow his general audience in the afternoon. He said he thought I might be interested in observing the response of the audience to the Pope's warmth and charismatic appeal.

It was exactly as Monsignor Cardinale had predicted.

More than one hundred people were waiting in the audience. When the Pope arrived, he quickly demonstrated a remarkable ability for making each person feel he was the recipient of individual Papal attention. The Pope sat in his high-backed, velvet-lined chair, his head resting lightly against the frame, and he spoke easily and informally about the meaning to him of Christmas. He related incidents drawn from his long pastoral life and he said that peace in our time had to be more than an aspiration; it had to be a reality. He blessed his audience, stepped down, and was assisted through the side door to his office.

Monsignor Cardinale and I followed the Pope into his oak-paneled study. No sooner had I entered the room than the Pope turned to me. "We have much to talk about," he said. "Just remember, I'm an ordinary man; I have two eyes, a nose—a very large nose—a mouth, two ears, and so forth. Even so, people sometimes remain rigid and uncommunicative when they talk to me. You must feel completely relaxed. We will talk as man to man." He smiled.

I handed him the letter conveying President Kennedy's concern and good wishes for his health. After he read the letter, I gave him the handwritten Christmas greeting from Nikita Khrushchev.

"The President is a wonderful man," Pope John said. "I have met some members of the family. They're all very fine people. The President is a splendid representative of the American people. When you return to the United States, I have something I want you to give to him from me for Christmas.

"It is nice of the President to be concerned about my health," he continued. "It is nice of Mr. Khrushchev, too, to send me a Christmas greeting. I get many messages these days from people who pray that my illness is without great pain. Pain is no foe of mine. I have memories. Wonderful memories. I have lived a long life, and I have much to look back upon. These memories give me great joy now and fill my life. There is really no room for the pain.

"There is so much to think back upon. When I was young I was an apostolic delegate in Bulgaria. I came to know and admire the Slavic peoples. I tried to study the Slavic languages, including Russian. I never became really proficient but I did learn to read the language to some extent. I am sorry I never pursued these studies. Do you speak the Russian language?"

"No," I said.

"A pity. You really ought to learn it. You are much younger than I. I am studying it. It wouldn't take you very long. A very important language. The Russian people, a very wonderful people. We must not give up on them because we do not like their political system. They have a deep spiritual heritage. This they have not lost. We can talk to them. Right now we have to talk to them. We must always try to speak to the good in people. Nothing can be lost by trying. Everything can be lost if men do not find some way to work together to save the peace. I am not afraid to talk to anyone about peace on earth. If Mr. Khrushchev were sitting right where you are sitting now, I don't think I would feel uneasy or awkward in talking to him."

I told Pope John that Chairman Khrushchev had expressed almost similar sentiments about him, pointing out the similarity of their peasant backgrounds, their upbringings in small villages, and their love of laughter.

"That is quite right." The Pope smiled. "When you live

close to the land, you have a real kinship for those who have done the same. As I say, I have never given up on the Russian people. Theirs is a deep spirituality that should never be overlooked. I don't think anything will change it."

Pope John listened intently as I gave the highlights of my report, beginning with an account of Mr. Khrushchev's response to the request for the release of Archbishop Slipyi.

"I have prayed for many years for the release of Archbishop Slipyi," he said. "Can you imagine what it must be like to be cut off for so many years from the kind of service you have prepared yourself to live, and from life itself? What is your impression? Do you think that the Archbishop will be released?"

I said I had no way of knowing. In any case, we would probably know before long. I then reported on the rest of the conversation with Mr. Khrushchev. When I completed the report, Pope John smiled and said, "Much depends now on keeping open and strengthening all possible lines of communications. During the terrible crisis over Cuba in October the possibility of a nuclear holocaust became very real. As you know, I asked the statesmen to exercise the greatest restraint and to do all that had to be done to reduce the terrible tension. My appeal was given prominent attention inside the Soviet Union. I was glad that this was so. This is a good sign."

His voice betrayed his fatigue and general sense of depletion, but he spoke with eagerness.

"I want to give you something," he said. He reached into a drawer and took out his personal medallion. "I hope you don't mind the absence of formal ceremony," he added. "It would please me to have you accept this little award for what you have done for Archbishop Slipyi."

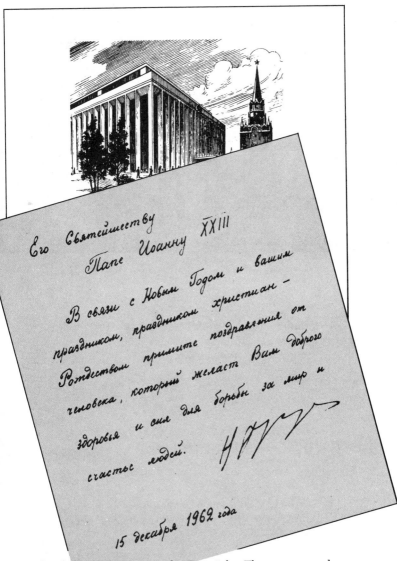

Его Святейшеству
Папе Иоанну XXIII

В связи с Новым Годом и вашим праздником, праздником христиан — Рождеством примите поздравления от человека, который желает Вам доброго здоровья и сил для борьбы за мир и счастье людей.

Н. Х.

15 декабря 1962 года

Khrushchev's Christmas card to Pope John. The message reads:
"To His Holiness Pope John XXIII   On the occasion of the Holy
Days of Christmas, please accept these greetings and
congratulations from a man who wishes you good health and
strength for your abiding quest for the peace and happiness
and well-being of all mankind.   N. Khrushchev   December 15, 1962"

Mr. Khrushchev uses his great bear coat
for his disappearing act as entertainment for some
young visitors from the U.S.

While escorting a guest through his vacation
place, Mr. Khrushchev provides a brief demonstration
of his badminton skills.

UPI-49

(TEST BAN)

MOSCOW--THE UNITED STATES, RUSSIA AND BRITAIN TODAY ENDED

HISTORIC TALKS THAT DIPLOMATIC SOURCES SAID SEALED FINAL AGREEMENT

TO END EAST-WEST NUCLEAR TESTS IN SPACE, IN THE ATMOSPHERE AND

UNDERWATER.

7/23-- TD1123AED

News wire from UPI sent by President Kennedy as a memento of the test-ban campaign. July 23, 1963.

I suggested that we wait until we learned whether the mission had been successful. If Archbishop Slipyi were released, I said half facetiously, perhaps Mr. Khrushchev should receive the medallion.

Pope John smiled.

"It's not appropriate for the Holy Father to bestow awards on heads of state," he said, "but I am going to give you two medallions. The first is for you. The second is for you, too, but with it I confer upon you the authority to award it with my blessing to anyone"—here he verbally underlined the word *anyone*—"you feel has deserved it."

I stood up to leave.

"World peace is mankind's greatest need," he said. "I am old but I will do what I can in the time I have."

Pope John walked with us to the door of his study and expressed renewed thanks for what had been done for Archbishop Slipyi.

On the way out to the car, Monsignor Cardinale emphasized that the Holy See was not attempting to arrogate to itself an unwelcome or unnatural role in its effort to reduce tensions between East and West. But there were so many elements of danger and so few elements of control that the Pope's efforts were essential. Even if these efforts were misconstrued, the Monsignor said, this was no warrant for inaction or absence of initiatives. The worst that could be said about Pope John was that he was taking Christianity literally.

After the Monsignor left me at the hotel I went for a long walk. I had much to think about. I could reflect that ours was an age which looked to physical motion for its spectacular achievements. The main articles of wonder in the modern world were men encased in metallic capsules spinning through outer space, or atoms pried open in order to release vast stores of energy, or streams of electrons following images of events happening thousands of miles away. But none of these miracles of motion would have the impression on history, I was convinced, of an eighty-one-year-old man, dying of cancer, using the Papacy to make the Church a powerful instrument in the cause of human unity and peace.

Just before I left for the airport the next morning, Monsignor Cardinale arrived at the hotel with a large object.

"I hope this won't complicate your baggage problems," he said. "It's a little Christmas present from the Pope for the President and Mrs. Kennedy. Here, let me show it to you. It's very easy to unwrap."

There emerged from the brown wrapping and tissue paper a silver icon about two feet long. Monsignor Cardinale said it was made in the ninth century A.D.

Later that afternoon, flying across the Atlantic, I pondered the vagaries of life that brought me to this particular point in space and time. I wondered what I would have thought many years earlier if, when I began my job at the *Saturday Review*, I had been told that in due course I would be in a propellerless plane above the ocean and that I would be delivering a medieval icon from the Vatican to the White House.

*E*ARLY IN JANUARY, 1963, I received a telephone call from Ambassador Dobrynin inviting me to lunch at the Soviet Embassy in Washington. I asked, as diplomatically as I could, whether the lunch was related in any way to my recent trip to Moscow.

The reply was in the affirmative.

Unlike most of the embassies, which are located in a diplomatic section several miles from the center of the city, the Soviet Embassy is on Sixteenth Street, N.W., in the heart of Washington, three blocks from the White House. The building is a converted old commodious mansion that still retains an atmosphere of old-world elegance with its curved marble stairways and high ceilings. A young male receptionist, Russian, presided over a large entrance foyer. Within several minutes of my arrival, an aide appeared and escorted me to the private apartment of the Ambassador on the top floor. As we stepped out of the elevator, the Ambassador came forward to introduce himself. Directly beyond him was his fashionably dressed and attractive wife, who welcomed me to their home. Ambassador Dobrynin was a tall, well-built man in his mid-fifties. His manner was open and informal. If he had any difficulty with the English language, I was unable to detect it.

We chatted a few minutes, then went into the dining

room. The table was set for two. Mrs. Dobrynin excused herself at this point. The Ambassador, after thanking me again for accepting his invitation, took out a piece of paper on which he had written detailed notes in a small, neat hand. He explained that Chairman Khrushchev had been in communication with him and these notes were the result.

"The Chairman sends his greetings," the Ambassador began, "and has asked me to report to you on developments since your visit several weeks ago. He is most pleased to be able to say to you that he has responded affirmatively to the specific suggestion concerning Archbishop Slipyi. As he promised, he looked into the matter. You will be pleased to know that the Archbishop is well and that he will be released in accordance with the suggestion.

"The Chairman has undertaken this action in the spirit of his conversation with you, in which the importance of strengthening the peace was recognized, and as a manifestation of his high regard for Pope John and the efforts being made by His Holiness in behalf of world peace."

The Ambassador then asked if I could undertake to inform him about the desired method of release. Was it the Vatican's wish that Archbishop Slipyi be brought to some neutral point such as Vienna and then turned over to a Vatican representative? He said he was certain that any suggestion made by the Vatican would be acceptable to Mr. Khrushchev.

I said that Father Morlion, who was due to meet with us in an hour, would convey the good news immediately to the Vatican officials and would obtain their advice on the method of Archbishop Slipyi's release. While I had no authority to speak for the Holy See, I knew I reflected the feelings of the Holy Father and those in the Vatican with whom I had discussed my trip to Moscow when I expressed heartfelt joy and gratitude at the news just conveyed.

The Ambassador said that he, personally, was pleased with these developments in view of what they portended for a future widening of the possibilities of peace. He said the Chairman was willing to pursue other suggestions I had conveyed regarding the religious situation in the Soviet Union. I replied that I had just been informed that Cardinal Bea, in charge of relations with non-Catholic Christian denominations, would be coming to the United States soon. The Cardinal would be pleased, I had been told, to have discussions with the Ambassador on these matters.

The Ambassador said he would welcome the opportunity to meet Cardinal Bea on the occasion of the latter's visit to New York. In any event, the Ambassador repeated his conviction that, while some events following Cuba had been somewhat disappointing, positive forces were now in motion.

This led to some personal reminiscences by the Ambassador, who said the prospects then were genuinely brighter than at any time since early 1960, just before the U-2 episode. He said he was close to the problem at that time. For five days after the U-2 plane had been shot down Mr. Khrushchev said nothing about it, hoping that the incident would not come to light. But the story got out in Istanbul. Even at that point, said the Ambassador, Mr. Khrushchev had assured his ministers that the President had known nothing about the flight. Then came the statement from Washington that the flights had been authorized by the President. This severely jolted the prospects for the Summit, but Mr. Khrushchev had decided to go to Paris nevertheless in the hope that the President might find some way of clarifying the situation. The Ambassador was with Mr. Khrushchev in Paris when a message came saying that the President would like to meet him privately to discuss the U-2 affair. According to the Ambassador, Mr. Khrushchev said he would be glad to meet with the President for this purpose but that the meeting never eventuated—and Mr. Khrushchev was at a loss to know the reason.

Paradoxically, I could report to the Ambassador that President Eisenhower had related virtually the same story to me when we discussed the matter. He had been eager to see Mr. Khrushchev for the purpose of saving the Summit Conference, if possible. Accordingly, he let it be known that he would like to meet Mr. Khrushchev privately, but he had never received a reply and he could not understand why none came.

We both could reflect that these were the vital intangibles that left a big question mark on history.

We then discussed American and Soviet relations in general. I expressed my personal view that one of the main elements in the difficulties between the two countries was that the Soviet government didn't know how to read American newspapers. I said this in all politeness, but I did not blunt the point. Admittedly, most of our newspapers were far from scholarly or detached, but one thing they had in common: they were not the official mouthpieces for the government. It was difficult for Soviet officials to recognize that what appeared in our papers was not a reflection of government policy. I had been given to understand that the Soviet decision to resume testing was based, in part at least, on their certainty, after reading our newspapers, that America had decided to test. Consequently, they felt they had no choice except to jump into it as quickly as possible. The actual fact was that President Kennedy had made a firm decision *against* the unilateral resumption of tests. It was only after the Soviets had tested that the President's hand was forced. Even then, he sought delay in order to achieve a test ban.

Still another instance was the Soviet decision to send missiles to Cuba. This decision was influenced, it seemed apparent, by the conclusion reached by Soviet leaders that the United States would not allow the failure of the first invasion attempt to stand, and that it was only a matter of time before a full-scale invasion would be under way. They reached this conclusion after reading our newspapers, some of which had called for full-scale invasion of Cuba. Here again the President had made a firm decision *against* invasion. When the build-up of Soviet missiles reached ominous proportions, the President had no choice other than to act as he did. In any event, if the Soviet government had been able to make a correct assessment of the President's position, there would have been no dangerous confrontation in Cuba between the United States and the U.S.S.R.

The Ambassador acknowledged that they might have made some faulty estimates but asked me to recognize, too, that what the newspapers said had a powerful effect on public opinion in the United States and that this public opinion was bound to have some influence on the decisions of the United States government.

"Even though what the newspapers say may not be wholly true," the Ambassador said, "it creates a general climate which causes most people—and not just the Soviet government—to believe that the worst is about to happen."

The Ambassador had made a good point, and I said so. But he had also opened up a question of the most profound importance; namely, where did public opinion really stand on major peace issues? I brought up a case in point. Early last year the matter of a government bond issue for the United Nations came before the Congress. Many newspapers were unenthusiastic about the bond issue, to say the least. Mail to senators and congressmen was running as high as fifteen to one against the United Nations in general and the bond issue in particular. The President was, naturally, apprehensive about what he considered a drift away from the United Nations. Friends of the United Nations in the United States got busy. With the cooperation of a few friendly congressmen, they undertook some scientific tests of public opinion in the United States and discovered that the sentiment in support of the United Nations and the bond issue was between eighty and ninety percent—and this held in almost every part of the country. The mail against the United Nations had been inspired by anti-United Nations groups. Consequently, when the President brought leaders of public opinion to Washington for the purpose of seeing how the "drift away from the United Nations" could be combatted, they were able to give him a factual report on the true condition of public opinion. He made immediate use of this material in his dealings with congressional leaders.

The Ambassador followed this account with the keenest interest, then said that even if his country were able to get accurate information about United States public opinion, they still had to make a determination about what the position of the United States government really was on any given issue.

This should be no mystery, I commented. The easiest way for Chairman Khrushchev to find out the position of the government on any given issue was to put the question to the President.

The Ambassador said that this was exactly what was now happening. No direct exchange had taken place between the time of the Cuban crisis and December 19, 1962, when the Chairman again wrote a personal letter on various matters on which the two countries ought now to seek specific agreement. Since then considerable progress had been made through such direct communication. In fact, the improved public tone was probably a direct product of such an exchange.

It was at this point that Father Morlion was announced. The Ambassador summarized for Father Morlion the message from Chairman Khrushchev. Father Morlion expressed his profound gratitude over the news about the liberation of Archbishop Slipyi and said that everything possible should be done toward creating the conditions that would enable the pledge of no publicity to be completely honored.

It was now almost 3:30 P.M., and we got up to leave. The Ambassador accompanied us to the main lobby. Father Morlion and I crossed the street, where we found a public telephone in the lobby of the National Headquarters of the United Automobile Workers.

We got through to Monsignor Cardinale in Rome without delay. He was overjoyed with the news about Archbishop Slipyi, said he would notify the Holy Father immediately, and get back to us as soon as possible about the method of transfer. Within a few hours we were able to inform Ambassador Dobrynin that the suggested place of transfer should be Vienna and that a Vatican representative would meet the Archbishop at the airport and proceed with him immediately to Rome.

The plan went off without incident. Archbishop Slipyi was visited at his place of detention by a top Communist official who said that arrangements had been completed for his release and that the Archbishop would be taken to Rome, where the Pope would greet him personally. He was informed that Mr. Khrushchev thought it might be nice if the Archbishop would wear his robes when he was reunited with the Pope. An aide told Archbishop Slipyi that Mr. Khrushchev had assigned his own tailor to the job. It was the first time that the most prestigious tailor in the Soviet Union had been given such an undertaking. (I later learned that the Archbishop could not accept the news as authentic. He thought at first it might be some sort of trick. It was only when he put on his church robes and looked at himself in the mirror that he realized his freedom might really be at hand.)

He was shown every care and courtesy on the trip to Vienna, where a Vatican representative was waiting for him. At Orte, not far from Rome, he was met by Monsignor Cardinale and escorted to a quiet religious rest home outside Rome. A few days later he was taken to the Vatican, where he was embraced by Pope John. After two days he returned to the rest home; his whereabouts were known to only a handful of people.

There was, however, an unfortunate occurrence that marred an otherwise happy ending to the project. Two days after the Archbishop's release I received a telephone call from Ambassador Dobrynin in Washington asking me if I had seen the afternoon newspapers. I said I had not. The Ambassador suggested that I do so. Then he read to me a news story under the following headline: *BISHOP TELLS OF RED TORTURE.*

He asked me if I would care to make any comment. It took me a moment or two to recover my composure. I said I had no direct knowledge of what had happened but I was absolutely certain that there had been no breach of faith. I said I would telephone the Vatican directly and find out what I could.

Monsignor Cardinale was profoundly shocked when I told him by telephone of the news break in the United States. He said Archbishop Slipyi had spoken to no newsman. He termed the story a pure concoction. He said the Vatican would set the record straight immediately. In particular, *Osservatore Romano* would carry a front-page statement quoting Pope John to the effect that the news stories about Archbishop Slipyi were without authority and would be repudiated by both Pope John and Archbishop Slipyi.

What troubled the Vatican officials most of all was that this incident might interfere with further attempts to bring about release of other churchmen imprisoned in Communist countries.

I telephoned Ambassador Dobrynin and informed him that the news stories were completely unauthorized and that the next issue of *Osservatore Romano* would make the correction on the authority of the Pope. Then I wrote to Chairman Khrushchev, emphasizing that at no point did the Archbishop see a news reporter. Hundreds of requests had been made for interviews or statements; all were refused. I also said in my letter that I had related to the Pope the account by Mr. Khrushchev of Marshal Tito's visit to Moscow and the erroneous radio report. And I referred to the point Mr. Khrushchev made that "some reporters just don't know how to handle good news."

Finally, I said I hoped Mr. Khrushchev felt I did the correct thing in reassuring the Pope that Mr. Khrushchev would not regard this news break an an evidence of bad faith. I referred to the text of a boxed item appearing on the front page of *Osservatore Romano:* "False news items were printed in the press and disseminated during the past few days concerning the case of Archbishop Slipyi. The Holy See and Archbishop Slipyi dissociate themselves completely from these reports."

There was no reply from Mr. Khrushchev to my letter of explanation, nor any further word from Ambassador Dobrynin. In fact, not until I saw Chairman Khrushchev three months later did I have any way of knowing whether the unfortunate incident involving Archbishop Slipyi would stand in the way of further efforts to release other churchmen or, in general, to improve the religious situation inside the Soviet Union.

Meanwhile, relations between the United States and U.S.S.R. seemed to be cooling again. Much of the positive momentum that had been building up had run out. The United States was insisting on eight annual inspections for a nuclear test-ban treaty. Khrushchev asserted it would be three inspections or nothing. He charged that the United States had reneged on its own proposal made by United States Ambassador Arthur Dean to U.S.S.R. Ambassador Vasili Kuznetsov at Geneva. A clamor was going up inside the United States for resuming nuclear atmospheric testing. It was argued that American scientists were on the verge of important breakthroughs in nuclear weaponry vital to the national security and needed to complete their research, which only testing could provide. It was also claimed that atmospheric testing was necessary in order to develop an anti-missile missile. Day by day the drums for resumption of testing beat louder and louder. And all the progress that had been made toward reducing tensions and slowing up the arms race began to fade.

Nikita Khrushchev's son-in-law, Aleksei I. Adzhubei, had visited Rome early in 1963 and had met with Pope John in a responsive visit to my own trip to Moscow. Unfortunately, it had been impossible to keep Adzhubei's visit out of the glare of the news, and the result was public misunderstanding concerning Pope John's purpose in seeing such an important Communist. There was mistaken speculation in the press that the Pope intended to extend formal recognition to a Communist society.

Adding to the general feeling of a downward pull at the time were reports from the Far East indicating that the Chinese leaders were stepping up their exploitation of the increased tension between East and West through new overtures to Moscow and appeals to the Communist parties throughout the world.

Father Morlion, who had just returned from a quick trip to Rome, told me of the mounting concern of Vatican officials over the sudden deterioration in the prospects for peace. The seriousness with which Pope John viewed the situation was indicated, he said, by the fact that he had decided to deliver a major encyclical on the peace and that it would be released within a few weeks. It was certain to be a document of historic importance. The advanced ideas I had heard Archbishop Dell'Acqua express during my December visit would be reflected, he believed, in the new encyclical.

Father Morlion said that the release of Archbishop Slipyi had brought feelings of profound thanksgiving to the Pope and the hierarchy in general. Cardinal Bea, a close friend of Archbishop Slipyi, was especially heartened by the release and wondered whether a similar effort ought to be made in behalf of Josyf Cardinal Beran, Archbishop of Prague, who had also been imprisoned for many years. Would I be available to undertake a return mission?

The December trip to Moscow had left me with a prolonged bout of dysentery and consequent feelings of debility. I suggested to Padre that perhaps someone else might be in a better position to carry out the venture. He said he had no desire to impose on my health but there was the greatest need for continuity.

While in Washington the following week I had lunch with Ralph Dungan. I asked if he felt any useful purpose might be served by a return trip to the Soviet Union. Ralph said he would put the question to Secretary of State Dean Rusk.

Several days later I returned to Washington in response to an invitation for lunch with the Secretary. Mr. Rusk said he was concerned about the impasse in the test-ban negotiations. He saw no prospect that the United States could reduce its demand for eight annual inspections. Khrushchev's contention that Ambassador Dean had made a specific offer of three inspections to Ambassador Kuznetsov was not justified. He had discussed this matter fully with Ambassador Dean and was convinced that the Soviet representative had misunderstood the offer. In any event, he said the test-ban treaty was in danger of being hopelessly bogged down. If I should see Mr. Khrushchev and the matter of the test-ban came up, the Secretary thought that I might be able to bear witness to the good faith of the United States in seeking an agreement. In particular, he thought I might suggest to Mr. Khrushchev that the test-ban negotiators on both sides should be authorized to proceed with all items other than the number of inspections. At least half a dozen important matters had to be negotiated. Mr. Rusk said I might suggest to Mr. Khrushchev that he and President Kennedy could meet after the negotiators completed all the other aspects of the treaty. They could then resolve the inspection matter. I was authorized to say that the President was fully confident that he and Mr. Khrushchev, working in good faith, could quickly agree on an acceptable number of inspections.

Following the lunch with Secretary Rusk, I saw Ambassador Dobrynin and asked if he would forward a request to Mr. Khrushchev for another meeting. A week later the Ambassador was on the telephone saying that Mr. Khrushchev would be pleased to see me in Moscow on April 12. I informed Ralph Dungan and Secretary Rusk.

The day before my departure for Moscow I received a telephone call from the White House. It was the President.

"About your trip," he said, "Dean Rusk has already spoken to you of our hope that we can get the test-ban unblocked. I have no doubt that Mr. Khrushchev is sincere in his belief that the United States reneged on its offer of three inspections. But he's wrong. He's basing his entire opinion on the Kuznetsov-Dean exchange. He's certain to tell you that Ambassador Kuznetsov is a good listener and a careful reporter. You can tell him that Ambassador Dean enjoys the same reputation and esteem with the President.

"You might also say that I have no disposition to argue about the Dean or Kuznetsov versions. It's possible both are right. I believe there's been an honest misunderstanding. See if you can't get Premier Khrushchev to accept the fact of an honest misunderstanding. There need be no question of veracity or honor on the part of either Kuznetsov or Dean, and the way can be cleared for a fresh start.

"Anyway, I'm sure you can support the fact that I am acting in good faith and that I genuinely want a test-ban treaty."

The President wished me luck and asked me to get in touch with him after I returned.

This time, mindful of the Chairman's admonition on my previous visit, I decided to take two of my daughters—Andrea, twenty-one, then a senior at college, and Candis, eighteen, a high-school senior.

We spent two days in Rome. As on the previous visit, I saw Monsignor Cardinale, Archbishop Dell'Acqua, Cardinal Bea, and Cardinal Tisserant. I was saddened to learn that Pope John's condition had worsened. More and more he had to carry on the duties of the Papacy from his bedside.

"The new encyclical has been completed; it will be a magnificent legacy," Monsignor Cardinale said. "It deals with the principal questions involved in war and peace. It will be called *Pacem in Terris*. We have great hopes for it. I can show it to you in a few minutes. It ought to be coming to my desk very soon from the Vatican print room. I have had a translation made into Russian. Perhaps you will want to take one with you to Mr. Khrushchev. He will be able to see it in advance of the general release on Friday."

A half-hour or so later a courier arrived with the first copies off the press of Pope John's new encyclical. I observed Monsignor Cardinale as he carefully removed the copies from the yellow folder. I had a sense that his own part in the preparation of the encyclical must have been an important one. He had an expression of genuine expectation as he turned the pages.

When the Russian translation of the encyclical arrived, Monsignor Cardinale reviewed with me the main points of the encyclical that might be pointed out to Chairman Khrushchev. He emphasized the passages dealing with the need to outlaw the weapons of mass destruction and the danger to the future generations of radioactive weapon explosions.

"We have great hopes," he repeated. "Very great hopes. It is a courageous document. It is the best of a very great Pope."

My next appointment was with Cardinal Bea and Father Schmidt. The Cardinal recounted for me the drama of Archbishop Slipyi's arrival in Rome. Even now, three months later, the Archbishop found it difficult to comprehend the reality of his freedom.

"How wonderful it would be if Archbishop Beran of Prague could also be freed," he said. "I know that it may be more difficult this time to persuade Chairman Khrushchev because of the terrible misstatements in the press after Archbishop Slipyi arrived in Rome, but I am sure you will do your best."

With this encouragement, and with an advance copy in Russian of *Pacem in Terris* in my briefcase, I took off for Moscow with my two daughters.

When our plane put down at the Moscow airport three hours later we were met by a number of Dartmouth conference colleagues, including Oleg Bykov and Alice Bobrysheva, and by a representative from Mr. Khrushchev's office who said that the meeting with the Chairman would take place at his retreat at Gagra on the Black Sea. We would fly down in the morning.

That night Alice Bobrysheva took the girls to the ballet while I had a long talk with Oleg Bykov about events in the Soviet Union since my last visit. He and another interpreter had been asked to accompany us to Gagra. The reason for the extra interpreter, he said, was that one of them could make notes while the other did the translating. The interpreter who was to join us was Boris Ivanov, who had been at the Andover meeting.

"Also," Oleg said, "it means that you'll have someone else to play chess with. He's really very sharp. One other thing. They've arranged for Alice Bobrysheva to come along so she can be of help to your girls for anything they would like to see or do."

Alice was a seasoned veteran of the Dartmouth conferences. Young, vivacious, attractive, she had endeared herself to members of both delegations with her ready availability for all sorts of chores—translations or logistics or letter writing, etc.

Our party of six took off the next morning for the jet flight to Sochi, where we were met by Chairman Khrushchev's personal representative, who had two cars waiting to drive us directly to the Chairman's retreat at Gagra, thirty miles away on the seacoast.

*T*HE ASPHALT ROAD FROM SOCHI TO GAGRA along
the edge of the Black Sea curls, climbs, and dives through the
rugged and verdant hills that drop down to the water. Automobiles proceed cautiously, not only because of the sudden
turns but because this road is a paradise for bicycle clubs.
Long skeins of cyclers, their backs bent low and their bodies
seemingly fused into the frames, come shooting at you in
endless swift files around the curves.

It was on this road, some thirty miles from the airport at
Sochi, that Nikita Khrushchev had his country retreat. The
house, large but not ostentatious, was set back from the road
behind a low wall in a grove of silver-streaked pine trees. The
place appeared to be lightly guarded. One man was posted at
the gate and waved us in when he recognized the driver.

As soon as we turned into the estate, I discerned a heavy-
set figure standing in the driveway in front of the house. It
was Mr. Khrushchev, patiently waiting to welcome us. He
was wearing a green-and-tan tweed cape and a large, gray,
unblocked fedora. I said I regretted that our various connec-
tions en route from Moscow had made us a half-hour late. He
replied that he would refer our apologies to the chef and suggested we proceed immediately to the luncheon table.

The dining room with its large glass doors looked out on

the sea. Mr. Khrushchev did the seating, explaining that Mrs. Khrushchev was in Moscow. Lunch was actually a full Russian dinner, with a vast assortment of appetizers, fish, soup, pancakes, veal, wines, cheeses, and pudding. The Chairman steered the table conversation; he had an anecdote to fit every course.

When I found some excuse for not going all the way with the substantial pourings of vodka, the Chairman told of the time he was in the company of some Georgians, who, in keeping with tradition, were drinking out of a massive wine bowl. As the bowl was passed, each man was expected to hold his own in terms of the duration and depth of a single gulp. The Chairman said he knew he was traveling in fast company but decided to take the bowl and the plunge nevertheless. "Served me right," he said. "I was sick for a week."

During tea he told the story of a frustrated tea drinker whose wife never gave him sugar for his tea. When away from home, he took out his resentment against his wife by thickening his tea with sugar to the point where he could hardly get it down. At home or away, therefore, the poor chap went through life without ever getting his tea just the way he liked it.

This is not to say that the luncheon was entirely without serious conversation. The Chairman asked whether we had observed all the flags on display along the roads and streets. This day, April 12, marked the anniversary of Yuri Gagarin's first flight into space.

"I was down here at the time and rushed up to Moscow to congratulate him," the Chairman said.

I said I understood from Evgeny Fedorov, one of the

leading scientists of the Soviet Union, that ordinary people would eventually be able to go up into space; in fact, that this was the way many people would want to take their vacations.

"It will be an interesting development, but it won't happen next week. Still, things are being simplified very rapidly and we hope before long to announce that we have trained a female cosmonaut."

"It would be interesting to get her reaction to a moon voyage," I observed.

"It would be interesting to get anyone's reaction to the moon," he replied. "Right now, however, it is difficult to predict when this will happen. My scientists tell me that they are ready right now, today, to put a man on the moon. But they can't assure me they can get him off and back home again. Of course I told them it would have to be a round trip. I understand now that the United States is very eager to be the first to do it. I say all the more power to you and good luck"—and he swept his arms in front of him in a polite gesture of stepping aside.

I asked the Chairman if he came to his Black Sea retreat in order to rest.

Not always, he replied. Sometimes he came here when he had important problems to think through or important speeches to write. He would walk through the pine grove or along the beach, and he would read and dictate. At such times he would shut off the telephone and tell the people in Moscow not to bother him.

"There are some things that can be done right only if you take the time they require," he said. "A chicken has to sit quietly for a certain time if she expects to lay an egg. If

I have something to hatch, I have to take the time to do it right. It is here that I thought through the problem of what to do about Stalin—whether to tell the people the truth about the man—especially about the tyrannical and irresponsible methods he used in personal dictatorship, or to perpetuate the myth of his greatness.

"Not that everything that happened in his regime was bad," the Chairman continued. "We made progress in a number of respects. But we were also held back in many ways because of the unbelievable irrationality and brutality of Stalin.

"It was not an easy decision to make, whether to tell the people the truth.

"I came down to this place and thought carefully about this problem and then decided to tell the Party Congress everything I knew. It was here that I also drew up the new economic program to increase production.

"It is very quiet here, as you can see," he said. "I have some visitors now and then; it's good to have a respite. Two weeks ago some Somali government officials were here for some brief talks. You are not the first Americans to visit this place. A few months ago Secretary Udall [*] was here. He made a fine impression on me. He said he had learned some things in observing our hydroelectric power developments. It is a big man who is willing to admit he can learn something from others.

"Also your John McCloy was here a few years ago. A very fine American and a gentleman. We went swimming together in the Black Sea. I think he enjoyed it. Then your Walter Lippmann was here a couple of years ago. He went swim-

[*] Stewart L. Udall, U.S. Secretary of the Interior.

ming, too. Also, your Eric Johnston was here. Americans make lively conversation.

"Now if you aren't too sleepy, we will walk around the grounds—if you would like to see the place."

I assured him I would, although I confess I got up somewhat heavily from the table.

Outside, we walked through the soft flooring of the grove of pine trees. The Chairman identified the trees as belonging to the rarest species of pine trees in the world. This was the only place, he said, where such trees had survived from their ancient beginnings. He was fond of these trees and had given many of them individual names but, like his grandchildren, there were so many of them that he was tempted to give them numbers instead.

The shaded walk soon led to a modern ranch-style structure on a hilltop. A glass wall fronted on the sea.

"This is my sport house," the Chairman explained. "First we will see the swimming pool."

We walked through a small indoor gymnasium and came upon the glass-enclosed pool. I judged it to be about thirty feet by seventy-five feet.

"The glass doors are electrically operated," Mr. Khrushchev said. "Here, I will show you."

He pressed a button and the giant doors began to retract.

Tongue in cheek, I told the Chairman that nowhere in the capitalist world had I seen a private swimming pool as magnificent as this.

The Chairman, with an equally straight face, consoled me, saying our society was still very young and that we would probably have one in due course.

When my daughters marveled at the swimming pool, the Chairman invited them to try it. They said they had not brought bathing suits.

"Don't let that worry you," the Chairman said. "Papa and I will look at the rest of the sport house, and then we will have our serious talk on the terrace and you will be all alone. You will have the pool to yourselves and will be undisturbed."

The girls decided they would like to complete the tour of the house first. The Chairman escorted us into the small gymnasium with its exercising equipment. When I asked what form of exercise he preferred, he pointed to the badminton racquets.

"I play badminton twice a day. Early morning and late afternoon. Then a swim and a rubdown."

I picked up one of the badminton racquets and bounced it against the flat of my hand.

"Do you happen to know anything about this little game?" he asked.

I confessed to some knowledge of the sport.

"Very well," he said, "we will have a go at it."

We picked up the racquets and started to play. The proprieties seemed to require that I hit the shuttlecock high and to his right side, just as I would if I were playing with one of my daughters.

After a minute or two of this kind of play the Chairman shook his head.

"*Nyet!*" he said. "That's not the way to play. My gymnasium instructor says that to play this game right you've got to hit the bird hard and fast and only a few inches above the net, like this—" Wham! And the bird came straight at my head.

Now that the ground rules were explicit, I no longer felt bound by excessive restraint. I was astounded at the speed of the Chairman's reflexes and his agility. He not only kept the bird in play but made it whistle as he rifled his shots.

When we stopped I observed that he was not winded or flushed. In a few days he would be sixty-nine. I thought of some newspaper stories I had read in Rome only four or five days earlier to the effect that he had had a heart attack or a stroke and that he had gone to the Black Sea to recuperate. Under the present circumstances these stories were less than convincing.

I asked the Chairman if he would permit me to take some photographs of him at his favorite sport. He assented readily and played badminton with my daughters while I operated the camera.

The tour of the sport house was resumed. Just outside the small gymnasium was the sun deck. Even on the coldest days, the Chairman said, he would come here to enjoy the sun. On these occasions he made ample use of a giant bear coat. He held it up; it was a massive garment indeed.

"Maybe the girls would like to see my disappearing act," he said.

He climbed into the coat, grinned, and went into a going, going, gone routine, finally sinking into the coat until he was completely out of sight.

Suddenly, there came a few growls from inside the encased mass. My daughters were delightfully terrified. Then suddenly the top flap flew open and he reappeared with a loud "Boo!" It was obvious that he had developed certain skills as a grandfather.

The Chairman put the bear coat down and said the time

had come for serious talk. "The girls are free to do what they wish. Papa and I will get down to business."

The girls went back to the pool. The Chairman, Oleg Bykov, Boris Ivanov, and I sat at a small table inside the glass-enclosed terrace.

On our previous visit he was relaxed, optimistic, confident. But now he seemed somewhat weighted down, even withdrawn. I couldn't be sure, but he seemed to be under considerable pressure.

Understandably so. Many things had happened to change the atmosphere since December. The Chinese had been exploiting the Russian missile withdrawal from Cuba, charging that Nikita Khrushchev was guilty of appeasing the imperialistic Americans. They claimed he had demonstrated his unfitness to lead the world revolutionary movement and that he had no real desire to overthrow or defeat the capitalist West, preferring to coexist with the very forces Marx and Lenin said must be violently overthrown. In return, Nikita Khrushchev had asserted that appeasement was in no way involved. He said the missiles had been installed in Cuba because of the possibility of an American invasion. Once the invasion threat was removed, there was no need to keep missiles there. At any rate, he had said that a nuclear holocaust over Cuba had been averted; this was the important thing. Anyone who knew anything about atomic weapons, he had declared, knew there was no alternative to peaceful coexistence. He had charged that the Chinese were absolutists who were attempting to use ideological dogma in places and situations where it didn't fit.

It was not at all surprising, therefore, that Mr. Khrushchev should seem preoccupied at Gagra. He had two critically important events coming up in rapid succession—the

Plenum of the Communist party and the confrontation with the Chinese. Either one called for important leadership decisions and actions. The combination of both would put him to the severest test since coming to office. He had come to Gagra before when he had serious problems to think through; this time the totality of his policies was involved.

We began our terrace discussion with matters that carried over from our talk of four months earlier.

I thanked the Chairman for his affirmative response to the request for Archbishop Slipyi's release.

Once again I expressed the regrets of Vatican officials at what had appeared to be a breach of faith in some of the news coverage that followed the Archbishop's release. I referred to the profound elation of Pope John at being reunited with Archbishop Slipyi.

The Chairman said he understood.

He then inquired about the health of Pope John, saying he had often thought of, and been inspired by, Pope John's desire to contribute to world peace in whatever time remained to him.

This seemed like a propitious moment to transmit to the Chairman the advance copy, translated into Russian, of Pope John's forthcoming encyclical, *Pacem in Terris*.

The Chairman said he was pleased to know about the encyclical. "Are there any parts of the encyclical," he asked, "that ought to be discussed now?"

This gave me the opportunity to call his attention to some of the key passages of *Pacem in Terris* dealing with the need to end the nuclear arms race and to regulate the affairs of nations in the human interest. The Chairman nodded frequently. He again praised Pope John for his service to world peace and said he would study the entire encyclical.

I brought up the matter of Archbishop Beran of Prague, who had been interned for some years. Cardinal Augustus Bea, of the Vatican had told me of his great concern for the Archbishop's health.

The Chairman said he was unfamiliar with the case of Archbishop Beran, and that this was a matter that concerned the Czechoslovak government.

Recognizing this, I said that Cardinal Bea was hopeful that the Chairman might be willing to use his good offices to explore the matter with Czech government officials.

The Chairman said he would take the matter under advisement.

The discussion then turned to the matter of a nuclear test ban. The Chairman had been quoted in news dispatches from Moscow as saying that the United States had not been acting in good faith on the matter of a nuclear test ban, reneging on its own proposals for three inspections, and that there was reason to doubt whether the United States really wanted a test ban. If he had been correctly quoted, I said, he might welcome reassurance on this score.

I had come to see him, I said, on no official mission; I was a private citizen. President Kennedy, knowing I was to see the Chairman, had asked me to try to clarify the Soviet misunderstanding of the American position on the test ban. If the Chairman construed the American position on inspections to mean that we actually did not want a treaty banning such testing, then that interpretation was in error.

The Chairman leaned forward in his chair. There was a perceptible tightening in his expression.

"If the United States really wanted a treaty, it could have had one," he said in measured tones. "If it wants one now it can have one. The United States said it wanted inspection. We don't believe inspections are really necessary. We think they are an excuse for espionage. Our scientists proved to me that new instrumentation makes it possible for you to detect any violations from outside our borders. But we wanted a treaty and the United States said we couldn't get one without inspections. So we agreed, only to have you change your position."

"There was a misunderstanding as to what our position really was," I said.

"A misunderstanding? How could there be a misunderstanding? Fedorov had a meeting with Wiesner [*] in Washington last October. Wiesner told him that the United States was ready to proceed on the basis of a few annual inspections. Ambassador Dean told Kuznetsov the same thing. Kuznetsov is a very meticulous reporter. He always tells me exactly what happened. How can there be a misunderstanding?"

I replied that the President had asked me to say that he had a high regard for Ambassador Kuznetsov and did not doubt for a moment that the Ambassador reported the conversation with Mr. Dean as he understood it. He also had a high regard for Ambassador Dean, who, like Ambassador Kuznetsov, had a reputation as a meticulously correct reporter. Rather than carry on a fruitless debate over the precise nature of the Kuznetsov-Dean conversation, the President was disposed to regard the matter as an honest misunderstanding; he felt a fresh start should be made. It would be a tragedy of the first magnitude, he believed, if a misunderstanding were al-

[*] Jerome B. Wiesner, Special Assistant to the President on science and technology.

lowed to get in the way of an agreement that both countries critically needed in their own self-interest and that would represent the first great step toward controlling the nuclear threat.

The Chairman shook his head.

"It is not just one conversation. As I told you, there was the talk between Wiesner and Fedorov. Also, our scientists came back from Cambridge, where they met with American scientists who said the same thing. How could there be a misunderstanding?"

With due respect, I ventured to suggest that an honest misunderstanding, under the circumstances, was possible and plausible. An American representative might urge the Soviet representative to revert to the previous Soviet position, which accepted three inspections, as the basis of an agreement. In so doing the American representative was suggesting what he considered to be the basis for negotiations that could lead to a prompt and fruitful resolution. The Soviet representative, however, might interpret the statement not as a basis for fruitful discussion but as the specific content of a treaty. The result was an honest misunderstanding.

In any event, I said, the President was acting in absolute good faith when he said that no misunderstanding, logical or otherwise, should obstruct so important an undertaking. I had firsthand evidence to offer on this point. A number of citizens' organizations had come together to develop public support for the President's position in favor of a nuclear test ban. In discussing this matter with the President, I had shown him the texts and layouts for a series of full-page newspaper advertisements calling for a test ban. The President was deeply interested in these materials and had constructive suggestions to make. It seemed to me inconceivable that he would have encouraged this public campaign if he had publicly advocated a test ban only for propaganda purposes, as the Soviet press had charged.

I had brought one of the advertisements with me and I held it up so that the Chairman could see it. The headline read:

>   (*In large type*)
>   We Can Kill
>   The Russians
>   360 Times Over
>   (*Then, in smaller type*)
>   The Russians Can
>   Kill Us Only
>   160 Times Over
>   (*Then, in very small type*)
>   We're Ahead,
>   Aren't We?

The Chairman stared hard at this advertisement while the text was translated for him. He lifted his hand.

"Your figures are all wrong," he said. "We're not that far behind. But, as the ad says, what difference does it make? Nuclear war is sheer madness. Now, back to our discussion: Your talk with the President has persuaded you of some things. Now let me tell you about the picture as we see it here. After Cuba there was a real chance for both the Soviet Union and the United States to take measures together that would advance the peace by easing tensions. The one area on which I thought we were closest to agreement was nuclear testing. And so I went before the Council of Ministers and said to them:

" 'We can have an agreement with the United States to

stop nuclear tests if we agree to three inspections. I know that these inspections are not necessary, and that the policing can be done adequately from outside our borders. But the American Congress has convinced itself that on-site inspection is necessary and the President cannot get a treaty through the Senate without it. Very well, then, let us accommodate the President.'

"The Council asked me if I was certain that we could have a treaty if we agreed to three inspections, and I told them yes. Finally, I persuaded them."

I thought of the earlier predictions by the Chinese that if the Soviet Union accepted the American proposal of three inspections, the Americans would renege and ask for six, and if Khrushchev agreed to six, the Americans would renege and ask for twelve. And so on, indefinitely. The Chinese position was that the Americans were interested neither in a nuclear test ban nor in coexistence in general. According to the Chinese, Khrushchev was naive in pursuing a policy of peaceful coexistence when the people he wanted to coexist with had no desire to coexist with him. The Chinese had quoted Marx and Lenin to support their view that war was inherent in the nature of capitalist imperialism and that the world would have to sustain a violent ordeal before capitalism could be cleared away and the triumph of world socialism assured.

Still another situation came to mind when the Chairman said it wasn't easy to get the ministers to agree. This had to do with the U-2 episode several years ago. At that time the Chairman had attempted to convince his Council that the American President had nothing to do with the U-2 and that he could therefore proceed with plans to meet with him at the imminent Summit Conference. Then the President announced he had authorized the U-2 flight.

These, at least, were some of the speculations that came to mind when the Chairman spoke of the reluctance of the Council to agree readily to his recommendation to accept inspection.

"People in the United States seem to think I am a dictator who can put into practice any policy I wish," the Chairman continued. "Not so. I've got to persuade before I can govern. Anyway, the Council of Ministers agreed to my urgent recommendation. Then I notified the United States I would accept three inspections. Back came the American rejection. They now wanted neither three inspections nor even six. They wanted eight. And so once again I was made to look foolish. But I can tell you this: it won't happen again."

"The President had no intention of humiliating you or making you look foolish before your Council," I said. "There is a genuine question in his mind concerning the adequacy of three inspections. Each year almost one hundred earth tremors or movements of varying magnitude occur within the vast land mass of the Soviet Union. Many of the seismograph markings caused by these movements are similar to the markings produced by underground nuclear explosions. Hence there is considerable feeling in the Senate that even eight inspections are minimal. In any event, the President would like to break the present impasse. He suggests that the negotiators at Geneva be instructed to proceed with the many questions apart from inspections that have yet to be worked out. These questions should represent no great difficulties, but they have to be resolved nevertheless. The President would like to hold the question of inspections for last, and then he and you would work out this problem together."

"Not practical or possible," Mr. Khrushchev said, again shaking his head. "For various reasons I cannot go to Washington and I would assume that the President right now has good reasons for not coming to Moscow. Where does this leave us?"

"It leaves you with the rest of the world in which to find a place," I suggested. "Vienna served the purpose once before. And if not Vienna, then another place. But even if no place can be found, then there are other forms of communication."

"You don't seem to understand what the situation is here," he said. "We cannot make another offer. I cannot go back to the Council. It is now up to the United States. Frankly, we feel we were misled. If we change our position at all, it will not be in the direction of making it more generous. It will be less generous. When I go up to Moscow next week I expect to serve notice that we will not consider ourselves bound by three inspections. If you can go from three to eight, we can go from three to zero."

He leaned forward in his chair.

"Now there's something else you ought to know," he said. "My atomic scientists and generals have been pressing me hard to allow them to carry on more nuclear tests. They believe that the security of our country requires that we develop new refinements in nuclear weapons. As you know, we have already successfully tested a 100-megaton bomb, but they want to follow this up with more variations. They say the United States has carried out seventy percent more tests than the Soviet Union and that the world will understand if we seek to reduce this gap. My scientists want a green light to go ahead; I think I may decide to give it to them."

For a moment or two I said nothing.

"Well?" he asked.

"You are looking at a depressed man," I said. "I came here for the purpose of bearing witness to the President's good faith. You have apparently placed little weight on this. Your final response is that you are probably going to resume atmospheric tests. If you do, I cannot imagine that the United States will stand still and let its lead dwindle. So we will test again, and you will test, and we will test, and so on. This destroys any possibility that other nations can be persuaded not to test. The poisons in the air will multiply. None of this adds either to American or Russian security.

"There is something else that occurs to me at this point," I continued. "Last summer President Kennedy was informed by a Soviet representative that missile bases were not being installed in Cuba. Perhaps it will be said that this was a misunderstanding. Under the circumstances, perhaps one misunderstanding can cancel out another."

Mr. Khrushchev looked at me severely.

"Very well," he said. "You want me to accept President Kennedy's good faith? All right, I accept President Kennedy's good faith. You want me to believe that the United States sincerely wants a treaty banning nuclear tests? All right, I believe the United States is sincere. You want me to set all misunderstandings aside and make a fresh start? All right, I agree to make a fresh start.

"Now," he said in unmistakably clipped tones, "let us forget everything that happened before. Forget all conversations involving Kuznetsov, Dean, Wiesner, Fedorov, and all the others. Now everyone will act in good faith and accept the good faith of everyone else. Very well. The Soviet Union now proposes to the United States a treaty to outlaw nuclear testing—underground, overground, in water, in space, every place. And we will give you something you don't really need. We will give you inspections inside our country to convince you we aren't really cheating. We make our offer; you accept it, and there's no more nuclear testing. Finished. If the President really wants a treaty, here it is."

He had left open the number of inspections. "The President has come down a great deal from the original twenty-two inspections," I said, "but he knows of no way he can come all the way down to three. The Senate would never accept it."

Mr. Khrushchev reached into the breast pocket of his blue suit and took out a "pull-out" watch—that is, a watch encased in a smooth metallic frame; when the two sides of this case are separated to show the time, the action also winds the springs. He toyed with the mechanism.

"We are repeating ourselves," he said. "Just so there is no mistake about it in your mind, let me say finally that I cannot and will not go back to the Council of Ministers and ask them to change our position in order to accommodate the United States again. Why am I always the one who must understand the difficulties of the other fellow? Maybe it's time for the other fellow to understand my position."

"This is precisely the position of the President," I said. "He does understand your position. That is why he suggests that misunderstandings be put to one side and that a fresh start be made."

Mr. Khrushchev sighed, then sat back in his chair.

"Very well," he said. "I agree. You can tell the President I accept his explanation of an honest misunderstanding and suggest that we get moving. But the next move is up to him."

I told the Chairman that what he had just said was entirely reasonable, and I thanked him.

The Chairman then asked if there was anything else I wanted to discuss.

I said there was, and I apologized for prolonging the meeting. I said that in lecturing before various groups in the United States, and in talking about the problems involved in a just and durable peace, I would constantly be confronted by people who would ask: "How can you talk about peace with the Soviet Union in view of the fact that Mr. Khrushchev has already declared war on us? He keeps saying he will bury us."

And so I asked Mr. Khrushchev how he would answer these questions.

"What I meant was, not that I will bury you but that history will bury you," he said somewhat testily. "Don't blame me if your capitalist system is doomed. I am not going to kill you. I have no intention of murdering two hundred million Americans. In fact, I will not even take part in the burial. The workers in your society will bury the system and they will be the pallbearers. Don't ask me when it is going to happen. It may not happen tomorrow or the day after. But it will happen. This is as certain as the rising sun."

I asked Mr. Khrushchev if he would be willing to consider evidence to the contrary.

"Please," he said.

I pointed out that Marx was unable to predict the profound changes that were to take place within the American economic structure. There was little resemblance between the capitalism Marx wrote about a century ago and the situation today. Instead of mass enslavement, the economic condition of the large masses of our people was vastly improved over what it was at the time of *Das Kapital*. We still had serious problems, of course. There was still a problem of waste—both with respect to our natural resources and our manpower; the nation was not yet making productive use of its black citizens. Even here, however, it would be a mistake to ignore important progress.

In any event, Marx had never fully anticipated the fluidity of a free society or the full significance of a considerable lack of acute class consciousness in the United States. Americans were productive and were improving their lot.

Mr. Khrushchev replied that if Marx were alive today he would not be dismayed by these developments but would say instead that all his predictions would come true. "I repeat," he said, "I have great admiration for the American people. Mark my word, when they become a socialist society, they will have the finest socialist society in the world. They are resourceful, energetic, intelligent, imaginative. What a wonderful thing this will be for them and for the world."

I told Mr. Khrushchev that the United States would be glad to have his good wishes but I thought it important to point out that notions of historical inevitability or determinism did not really fit American history or the American character. Peaceful coexistence, as I understood it, meant that each state could hold to its institutions and there could still be peace.

"*Harrasho,*" he said. ["Just right."]

I knew I had prolonged our talk far beyond any reasonable limits. There was, however, one additional assignment I had been asked to carry out. Rex Stout, the mystery-story writer and president of the Authors' League, had empowered me to represent the League in seeking some solution to the copyright tangle with the Soviet Union. For the past ten years various attempts had been made to persuade the Soviet Union to respect American copyright. The main countering argument on the Soviet side was that the Russians read many times more of our books than we did of theirs and that we were therefore proposing an unfair balance of literary trade. Another problem had to do with retroactivity. Publication of American books without authorization in the Soviet Union had been going on a long time; how far back in time would the Soviet liability go?

I told Mr. Khrushchev that the Authors' League authorized me to propose that any copyright agreement would be free of past liability. As of January 1, 1964, say, each country would honor copyright restrictions and seek permission for any literary works originating in the other country. No retroactive payments would be required, although continued publication and distribution of books issued before January 1, 1964, would, of course, be covered by the new agreement.

Mr. Khrushchev shook his head.

"What kind of a deal is this?" he asked. "You get all the benefits, and what do we get? We publish maybe millions of copies of books by American authors. We read Hemingway, Faulkner, Mitchell Wilson, Jack London, Mark Twain, Sinclair Lewis, and many others. And how many of our writers do your people read? A few of the classical ones but hardly any of the contemporary ones. We are a nation of book readers. You are a nation of television watchers and comic-book buyers. How can you propose a deal when you are not in a position to offer anything?"

I told the Chairman that there was increasing interest among Americans in contemporary Soviet authors, but even if this were not the case it was hardly relevant, one way or the other. American writers were entitled to payment for the use of their words. They also had the right to decide whether their books should be reprinted. If we could agree on this principle, then we could talk about the entirely separate matter of increasing the availability of Soviet books in the United States.

"You may think these are separate matters but we do not. I see no chance right now for a copyright agreement," he said. "But we would be glad to talk to you about developing something approaching parity in our literary exchange. Once that is done, we can consider the copyright problem. But, as I say, your country has a long way to go before you can equal ours in the matter of book reading. You know Tolstoy, Dostoevski, Gorki, and one or two others from the old Russia, but you know very little about our living writers—and we have some good ones. And even what you know of Tolstoy is badly corrupted. When I was in the United States several years ago I saw a version of *War and Peace* in comic-book form. It was made into a story of terror, wild sex, and brutality. How can you expect poor Tolstoy to rest in his grave with nonsense like that going on?"

I had no defense to offer for the offensive edition of Tolstoy and said so. But the context in which the Chairman made his remarks was incorrect. He said our people did not read books. An error. Last year more than three hundred million books were sold in the United States. This did not include the comic books or the cheapies he had spoken about. Was he familiar with the large number of serious books now being published and sold in inexpensive paperback form? Again this didn't excuse the cheap books, but at least the situation was somewhat different from the way he had understood it.

"You are to be congratulated," said the Chairman. "Now if only you can do something about your television, you will be making some real progress. Frankly, I could hardly believe my eyes when I was in the United States, the kind of things you showed on television. If the sadism and violence you show are at all representative of the kind of life you have in America, God help you! All the killing and beatings and cheating and swearing and wife stealing and immorality! A nation can't help being judged by the things it is interested in.

"But what is most surprising to me," he continued, "is that you apparently have no idea of the kind of harm this is doing to your children. They sit in front of the TV sets for hours at a time and take it all in. What kind of food is this for tender young minds? And you wonder why you have a juvenile delinquency problem. Surely your capitalists, who put on these TV programs must have some conscience and can be persuaded not to make money out of deforming children's minds. And if they can't, why can't your society do something about it? Capitalism isn't just an unjust economic system. It's a way of life that leads to a corruption of important values. Television is only one example."

Once again I asked the Chairman if he would be willing to consider some contrary evidence. And once again he said, "Please."

First, concerning television. I said he made it appear that there was almost no concern in the United States about this problem. Had it not been called to his attention that the chairman of the Federal Communications Commission had attacked irresponsible programming on television? Was he unaware of the various citizens' groups that had been organized to combat harmful TV? Had he overlooked the numerous articles in the press on the subject? In any event, two important facts had to be stressed. The first was that the American people had freedom of choice. They were not confined to a government station; in many cities they had four or five or more channels. They also had freedom of choice over a wide range of programs dealing with serious music, good films, news reports, and public affairs. They could watch debates over government policy, in which the government itself would be seriously criticized. No penalties would be attached to such criticism. Moreover, some of our large capitalist companies would sponsor important music events and other similar high-quality programs.

Apart from all this, a major development in American television had probably escaped his attention. I referred to the fact that there were now one hundred educational television stations across the United States. These stations were free of what he called "trash." The fact that they were called "educational" TV stations did not mean that they were used just for institutional purposes. They provided general programs of genuine merit, combining public education with high-level entertainment. The national educational organization that supplied many of the programs for these stations was financed by a foundation that obtained its money through the sale of automobiles. Locally, the stations were supported on a community basis, and many capitalistic enterprises contributed to the upkeep of these local stations without commercial announcements or expectation of profit.

The Chairman said he hadn't intended to offend me or arouse me and was glad to learn about all these promising developments in the United States. And he was especially glad to know that so many people were eager to do something about the awful trash on television. He wished them all the luck in the world and said they would probably need it.

Apart from educational TV, there were other impressive indications, I said, of a healthy growth in the creative and cultural life of America. The fact that one child out of three was taking instrumental lessons in music; the fact that millions of people attended concerts each year—more, in fact, than went to baseball games; the fact that American colleges and universities were now undergoing the greatest expansion in their history—all these facts might indicate that the United States was not so backward or underdeveloped culturally as his earlier remarks would seem to indicate. The Americans were putting their freedom to good use.

The Chairman said he applauded these developments and could only say that if the American people had done this well despite their present system, just think of the kind of progress they would make when they turned socialist. And once again he looked at his pull-out watch.

It was late in the afternoon and the sunlight was waning. I was eager to put a final question to the Chairman. I asked him whether he was discouraged in his stated policy of peaceful relations with the United States and the West in general.

The Chairman said he wanted to believe that the terrible drift to war could be ended and that the two most powerful countries in the world could find some way to live in peace, but that the next move was up to the United States. He looked to President Kennedy, for whom he had high regard, to take the next step. Then he asked me to convey his greetings to the President and Mrs. Kennedy. He asked that I also convey to Pope John his concern over the latter's health and his best wishes.

"I appreciate very much the Pope's personal medallion you sent me," he said. "I keep it on my desk at all times. When Party functionaries come to see me, I play with it rather ostentatiously. If they don't ask me what it is right away, I continue to let it get in the way of the conversation, even allowing it to slip through my fingers and to fall on the floor, so that they have to watch out for their toes. Inevitably, I am asked to explain this large engraved disc. 'Oh,' I say, 'it's only a medal from the Pope. . . .' "

He enjoyed his story. This seemed to me to be a good time to bring up something that I had hoped I might ask ever since I knew I would be coming to see him a second time.

"May I change the subject?" I queried.

"Of course."

"I know I'm taking advantage of my being here to ask you something you might rather not discuss."

He smiled. "With a preamble like that, I can hardly wait to hear your question."

"What would you say your principal achievement has been during your years in office?"

"Could I talk about two achievements and not just one?" he replied. "The first was telling the people the truth about Stalin. There was a chance, I thought, that if we understood what really happened, it might not happen again. Anyway, we could not go forward as a nation unless we got the poison of Stalin out of our system. He did some good things, to be sure, and I have acknowledged them. But he was an insane tyrant and he held back our country for many years. The second thing I am proud of is that it will not be necessary for my successors to shoot me when the time comes for me to leave this office. So my second achievement is very much like the first: I think I have made it possible to have an orderly transfer of authority—not just for the men who replace me but for the men who will replace them."

"How do you feel about the fact that you opened up relationships with other nations? Do you feel your discussions with President Eisenhower and President Kennedy represent a historic step in a new foreign policy of reconciliation with the West?"

"You're making some good suggestions for my memoirs," he said. "One of Stalin's great mistakes was to isolate the Soviet Union from the rest of the world. We need friends. We have mutual interests with the United States. These two great countries would be very stupid if they ignored these mutual interests. They also have serious differences. But no one need worry that these differences will be glossed over. There are people in each country who make a career out of the differences. But someone has to speak also of the serious mutual interests. I have tried to talk about them."

My daughters and their escort approached the terrace. We stood up to greet them. The girls had been swimming in the Chairman's pool (later they told me that they had used Mr. Khrushchev's trunks—"they ballooned out like life preservers"). Then we all walked back to the main house. I thanked the Chairman for being so generous with his time.

When we said good-bye, I remember thinking at the time that the person of Nikita Khrushchev didn't fit the image one might have of a powerful political leader. There was nothing exalted about the man. He was a lonesome figure who gave the impression of being gregarious. He was a man who obviously managed to make time in his own life for sustained and sequential thought. Yet he was identified in the public mind, and for good reason, as a man who charged about with a great deal of impulsiveness and flurry. He never attempted to conceal his peasant background and this was his great strength; yet he didn't hesitate to wear expensive silk shirts and gold cuff links. He was supposed to be crude, yet I had seen that he was capable of gentility, kindliness, and great courtesy. I could reflect that if one of the persistent characteristics of prominent leaders in history was a large assortment of paradoxes—and if it is true that a man comes to life in his paradoxes—then it was clear that Nikita Khrushchev was a full-sized leader and one of the substantial figures of the twentieth century.

ON April 22, 1963, I went to Washington to report to the President. I was ushered from the reception room into the Cabinet Room facing the White House lawn.

Looking out through the colonnade doors while I waited for my appointment, I could see workmen swarming all over the lawn. They were arranging benches in front of a performance stand. A dozen or so teen-age boys and girls with official armbands were assisting with the arrangements.

After twenty minutes the door leading to the President's office opened and a group of educators emerged, led by Commissioner of Education Francis Keppel. Keppel explained that he had just introduced the teacher of the year to the President as part of an annual ceremony. As Keppel was finishing his account, the President came up behind him, grinned, and said he was still awed by teachers. Then the President took me into his office and told me to sit on the large sofa. At that moment an attractive young woman opened the door, waved at the President, saying all she wanted to do was to say hello.

"Come on in, Eunice," the President said. "I want you to meet someone. This is my sister, Eunice Shriver. Eunice, I want you to meet Norman Cousins."

"Hello, Mr. Cousins," she said. "My congratulations on the honor."

I took her hand, then looked at her quizzically.

"Congratulations?" asked the President. "What honor?"

"For being selected teacher of the year."

"Eunice," said the President. "Mr. Cousins is not the teacher of the year. He's the editor of the *Saturday Review*."

Mrs. Shriver recovered quickly.

"Well, they *did* tell me the teacher of the year was inside," she said.

"That was ten minutes ago." The President smiled.

"Well, anyway, Sargent always reads your magazine and quotes it to me," she said. Then she kissed the President and left.

"A lot of activity going on today," the President said. "All that bustle on the lawn outside is for a high-school musicale that Jackie arranged. The kids are arriving by the busloads right now. This is Jackie's department, but Jackie is away. I understand you had a good meeting with Chairman Khrushchev. It was at his place on the Black Sea, wasn't it?"

I nodded.

"Is it as nice as they say?" he asked.

I told the President about the grounds, the rare trees, the main house, the swimming pool with its electrically operated retractable doors, and the carpeted gymnasium. I also told him of my ten-minute badminton game with the Chairman.

"I had heard it was a pretty plush place," the President said. "How did you do at badminton? I trust you didn't win."

"It never really came to that. Khrushchev gave me instructions. He was swatting the bird and wanted me to do the same."

"Sounds as though he's in good condition."

I told the President I thought Khrushchev's condition was

quite remarkable, considering the number of vodkas he had consumed at lunch. The exertion at badminton had left him neither winded nor flushed.

Then the President asked about the substance of the conversation with Khrushchev—in particular, how the Chairman had reacted to the message the President asked me to give about the fact that the United States genuinely wanted a test-ban treaty and that the President believed there had been an honest misunderstanding over the number of inspections.

I gave the President a full report of the long meeting with Chairman Khrushchev. I used complete quotations from my notes, as in the case of the Chairman's solemn assurances to the Council of Ministers that the United States was ready to proceed with a treaty on the basis of three inspections. I told of the awkward situation Khrushchev said he faced when it developed that these reassurances were worthless. I also reported the Chairman's remark that once again he was made to look like a fool, but that it would not happen again. I emphasized that a large part of the conversation was directed to the misunderstanding over the number of inspections.

The President sat quietly in his rocker. His eyes were fixed directly on me as I spoke. Then he said:

"You know, the more I learn about this business, the more I learn how difficult it is to communicate on the really important matters. Look at General de Gaulle. He's one of our allies. If we can't communicate with him and get him to understand things, we shouldn't be surprised at our difficulty with Khrushchev."

I spoke of the situation inside the Soviet Union and the pressure on Mr. Khrushchev to adopt a hard line.

The President said: "One of the ironic things about this

entire situation is that Mr. Khrushchev and I occupy approximately the same political positions inside our governments. He would like to prevent a nuclear war but is under severe pressure from his hard-line crowd, which interprets every move in that direction as appeasement. I've got similar problems. Meanwhile, the lack of progress in reaching agreements between our two countries gives strength to the hard-line boys in both, with the result that the hard-liners in the Soviet Union and the United States feed on one another, each using the actions of the other to justify its own position."

Then the President spoke of the problems that grew out of some inaccurate stories in the press. These stories were picked up in the Soviet Union and used to justify a hard-line position there. Meanwhile, we counter-reacted to their fulminations.

All in all, he said, it is a complex problem but one we had to find some way to solve.

In any case, the President said he was pleased when I told him that Khrushchev had finally said he was willing to make a fresh start on the test-ban negotiations and eagerly awaited some official initiative in that direction by the United States.

"This looks like a job made to order for Averell Harriman," the President said. "As for Khrushchev's difficulties with his Council of Ministers, I think I understand his personal situation. He's on the spot, but I don't see how we can cut down on inspections. We'd never get it through the Senate. As it is, we'll have a real battle on our hands to get a treaty through the Senate even if the Russians agreed to everything we asked. Maybe we can find some way around the impasse on underground testing. Dean Rusk isn't in Washington today but he'll be back next week. Perhaps you can arrange to come back and see him then. If you're free this afternoon, you should see Tommy Thompson and Bill Foster."

I told the President I was completely at his disposal.

The President picked up his telephone and asked Mrs. Lincoln to put through calls to Ambassador Llewellyn E. Thompson and William C. Foster. When each man came on, the President summarized the background of my trip and made an afternoon appointment.

Then the President turned to me and said the account of Mr. Khrushchev's difficulties in bringing along the Soviet leadership on the test-ban issue confirmed his own belief that Khrushchev, far from being in a position of absolute power, had plenty of problems on his hands in getting the support he needed for his program.

I told the President that the evidence seemed to suggest that a political crisis was developing inside the Soviet Union, and that the test ban might be something of a pivotal issue. Khrushchev was staking a great deal on his ability to show results for his make-peace policy with the United States. That policy would make possible an increase in consumer goods and services and an upgrading in the Russian standard of living. Conversely, the failure of that policy would require a high level of non-consumer production for weapons, and would strengthen the position of the anti-West hard-liners inside the Soviet Union and the Communist world in general. The situation was clearly one in which the Chinese couldn't go along with Khrushchev's coexistence policy without upsetting a whole host of other factors that affected their position in the world. The Chinese invoked Marxist-Leninist dogma about the inevitability of war in the nature of capitalist society—not just because they were allowing themselves to be manipulated by theory but because they were using theory to bolster a policy decided upon for other reasons.

In any event, the test-ban issue had become something of a watershed inside the Communist world. The failure to achieve it would be hailed by the Chinese and their supporters as proof of the unrealistic nature of Khrushchev's policy.

There was some indication, I told the President, that the Chinese had already written off the chances for successful completion of a test-ban treaty. The present impasse, now being exploited in their propaganda, had apparently convinced them there would be no further negotiations. They apparently believed the Soviet leaders were getting ready to make a public acknowledgment that the projected treaty was hopeless. The Chinese apparently also expected that the inevitable result would be improved rapport between Peking and Moscow. They were already moving forward on this assumption: they had asked the Soviet leaders to accept the visit of a high-level delegation later in the spring, perhaps in June.

The President said he recognized the complexities of Chinese-Russian relations. Some things were beyond our reach or our power. But one thing that might be within reach was improved American-Soviet relations.

In that case, I said, perhaps what was needed was a breath-taking new approach toward the Russian people, calling for an end to the cold war and a fresh start in American-Russian relationships. Such an approach might recognize the implications of a world that had become a single unit, however disorganized; it might recognize, too, that the old animosities could become the fuse of a holocaust.

The President lit a thin cigar and said he would like to

think about it. He asked me to prepare a memorandum for him on the subject.

Then the President questioned me about other matters discussed with Mr. Khrushchev.

I told him about the misunderstanding that had to be cleared away as the result of the unfortunate news break at the time of Achbishop Slipyi's release. I said that Khrushchev's own account at the December meeting, of the erroneous and embarrassing news report about Marshal Tito's speech and his apprehension over the possible misunderstanding it would cause, gave me a peg on which to hang our own explanation.

"Just more examples of what I spoke about earlier when I referred to the difficulties of communication," he said. "Did you satisfy him on the Slipyi case?"

"I don't know. I hope so. It was a little sticky because I was asked by Vatican officials to try to negotiate the release, too, of Archbishop Beran, head of the Catholic Church in Czechoslovakia."

"Did he agree?"

I said that Khrushchev had intimated it might be more difficult to arrange than the Slipyi release because Archbishop Beran was outside his jurisdiction. Khrushchev was not upset by this request and said he would see what could be done. [*] I told the President I was relieved that the Chairman apparently had not allowed the Slipyi case to interfere with further steps.

The President commented that the meeting sounded most productive. I responded that I was afraid I had tried the Chairman's patience; I spoke of the Chairman's critical references to American society—in particular his notions concerning our

[*] Archbishop Beran was released several weeks later, in May 1963.

readiness for the mortician's slab. I also spoke of the Chairman's comments on the state of our culture, then related the statistics I had cited in reply.

"If you have any statistics about the state of American culture, I wish you would give them to me," the President said. "In about half an hour I've got to go out on that lawn and say something to those high-school kids. Jackie's in Florida and I've got to look after her music party. What did you tell the Chairman about music?"

"Just that more Americans go to concerts each year than go to baseball games. I told the Chairman about the large number of youngsters who study instrumental music. I told him that more than three hundred million books were published in the United States last year. I told him something about one hundred American educational television stations and what they were doing to bring good music, drama, and serious discussion to the American people.

"Just what I need today," the President said. "I don't have to talk more than ten minutes. Do you think you could jot down a few notes for my talk at one o'clock?"

I looked at my watch. It was twelve thirty-five. I told the President I would do my best to do a draft in twenty minutes for a ten-minute talk.

"Mrs. Lincoln will be here to help you in any way you wish," he said. "She can take dictation or make telephone calls for you. I'm going to see some of my staff in their offices. I like to drop in on them before they drop in on me. Be sure to telephone me if you want more background about the talk."

He moved toward the door, then turned.

"I just want to thank you again for bringing the icon from the Pope. My experts tell me it's a fine specimen."

Very briefly, I told the President about the incident concerning Mr. Khrushchev and the Pope's medallion. He especially enjoyed the Premier's efforts to call attention to the Papal present.

"You know," he said, still smiling, "that's one advantage a Communist leader has over an American President, especially a Catholic. Khrushchev can go out of his way to boast about receiving a gift from the Pope. I've got to be careful about these things."

After the President left, I told Mrs. Lincoln I might be able to work more quickly on the typewriter than I could by dictating. I telephoned my office in New York and asked Mary Harvey for top-speed service in checking the number of books published in the United States in 1962, the number of educational television stations, the number of students taking instrumental music lessons, the comparative figures on attendance at concerts and baseball games, and the number of students studying art. Apologetically, I gave Mrs. Harvey six minutes to get back to me on the telephone.

Mrs. Lincoln set me up at a typewriter in the President's office and I went to work on the draft, leaving blank spaces for the figures.

At 12:48 P.M. Mrs. Harvey was on the telephone with all the required information. At 12:51 P.M. the President was on the telephone again. He wanted to be sure I didn't leave before having lunch in the White House dining room. He said his afternoon appointments began immediately after his meeting with the students on the lawn, but that Mrs. Lincoln would probably find some way of getting a glass of milk and a sandwich into him between appointments. He said he would call Ralph Dungan and ask him to have lunch with me. Before hanging up he reminded me to be sure to telephone him if anything came up during my afternoon appointments that I wanted to discuss with him directly.

I finished the notes for the President's talk to the youngsters. Ralph Dungan picked me up and took me to the White House dining room, located in the lower level in the executive offices. There were a half-dozen or so tables. The menu was mimeographed. The fare was simple: two or three appetizers, soup, several sandwiches, macaroni and cheese, chopped sirloin steak, salmon soufflé, the standard desserts.

During soup, Ken O'Donnell, the President's appointment secretary, came over to our table.

"Now I've seen everything," he said. "The President had seven minutes before he was due to speak. He raced down to the White House pool, tore off his clothes, dove in, and swam with one hand. In the other hand he was studying the draft of the talk you gave him for the kids."

My first appointment that afternoon was with Llewellyn E. Thompson, former ambassador to the Soviet Union and now special consultant on Soviet affairs. "Tommy," as he was known in the diplomatic corps, had a wide knowledge of Russian history, culture, traditions, and psychology. He was extremely popular with the Russian people. While in the Soviet Union he would occasionally appear on television and would speak in Russian. He was open and direct and had won himself a position of high esteem among Soviet officials.

In talking to Tommy Thompson I found substantial verification for the key conclusions I had reached on my latest trip to Moscow. He was able to fill in for me background on Chairman Khrushchev's difficulties with the Council of Min-

isters. He felt that a special effort should be made to keep Soviet policy from tightening up again. In particular, he believed
that the test-ban treaty deserved another try.

My next meeting with William C. Foster, chief of the
U.S. Arms Control and Disarmament Agency, and with his
deputy, Adrian S. Fisher, was similarly productive. They believed that Ambassador Kuznetsov and Evgeny Federov had
jumped to unwarranted conclusions concerning the American
position, but did not doubt that the Russians sincerely believed
that we had backed off from what they considered to be a
bona fide offer of a treaty based on three inspections. In any
event, they felt that a new formula had to be found before a
new hard-line developed inside the Soviet Union, one aspect
of which concerned relations between the U.S.S.R. and China.

I left Washington that night with a sense that, although
the general news was bleak, the policy-makers in our government were determined to find some way of making a fresh
start.

*T*HE NEXT MORNING I wrote a memorandum for the President, in accordance with his request, summarizing the main points of our discussion and giving special emphasis to the projected visit to Moscow of the Chinese delegation early in June. There was no immediate acknowledgment, but two weeks later I received a telephone call from Theodore Sorensen asking me to come to Washington.

Sorensen's office, commodious and sedate, adjoined the President's. I had met Ted Sorensen three or four times previously. I knew him to be precise in his judgments and straightforward in manner. There was nothing ornamental about his ideas or the way he expressed himself. The range and incisiveness of his intelligence were enormously impressive.

Sorensen said the President had given him the memorandum in which I had suggested a dramatic peace offer to the Russians that might unblock the nuclear test-ban impasse before the Chinese Communist delegation arrived in Moscow. He said he felt the Soviet-Chinese rift was so deep that it would probably be unaffected even by a breakdown in the test-ban negotiations. He also felt that if the Soviet leaders were ready for agreements leading to a real improvement of the chances for peace, the approaches to such agreements ought to be pursued.

He said the President had spoken to him about the possibility of my sending in some ideas for the text of a commencement talk at American University in Washington, June 10, 1963. The President felt that some of the points I had made in my conversation with him and in my memorandum might be incorporated in the draft. I thanked Ted Sorensen and said I would send some notes within a few days.

Even before the President's June 10 speech at American University in Washington, D.C., there was a highly positive development. On May 27, 1963, Senator Thomas E. Dodd of Connecticut, and Senator Hubert H. Humphrey of Minnesota, introduced a Senate resolution calling for a nuclear test-ban treaty that would by-pass the inspection issue by limiting itself to atmospheric testing. Since the massive explosions had to be carried out above ground, and since one of the major reasons for a test-ban was the need to prevent further radioactive contamination of air, earth, food, and human bone and tissue, the Dodd-Humphrey resolution seemed an adroit way of breaking the deadlock over inspection. The idea of a partial test-ban, of course, had come up at various times during the long history of the negotiations. The Eisenhower Administration raised it as a possibility in 1959, and the Kennedy Administration suggested it in the summer of 1962. But the fact that a substantial number of Senators, including several who had been firmly against any ban, were now on record in support of a partial treaty, may well have been a vital factor in the long struggle against testing of nuclear weapons. Thirty-four Senators joined Dodd and Humphrey in signing the resolution. For the first time, President Kennedy could feel some momentum behind him on the test-ban fight.

The big question, of course, was whether Khrushchev

would agree to a limited treaty. Would he hold that the cut-down version was worse than nothing at all? A great deal depended on the general atmosphere in which a limited treaty would be pursued. The President's talk on June 10 could have a profound effect on that atmosphere. It could create a favorable context for the consideration of such a treaty.

Ambassador Averell Harriman brought back word from Moscow confirming that Premier Khrushchev responded favorably to the idea of a fresh start on the test-ban treaty. Harriman, whose experience in dealing with Soviet leaders at the highest level was second to none, had put before Khrushchev the argument that a test-ban treaty could represent a major breakthrough in East-West relationships.

Then came President Kennedy's American University speech. The timing was most auspicious. The Chinese delegation was just arriving in Moscow. Coincident with the arrival of the delegation, Peking addressed an open letter to the Russian people. It was obvious from the tone and content of the letter that the Chinese were exulting in what they had assumed was the death of the test-ban negotiations. They apparently dismissed the formula recommended in the Dodd-Humphrey proposals as unacceptable to Premier Khrushchev, who was on record as having declared much earlier that he wanted a complete treaty or nothing at all.

The Chinese open letter seemed to reflect China's confident belief that the Khrushchev coexistence policy would either have to be scrapped and a new initiative toward Peking adopted, or that Khrushchev himself would have to give way to a successor whose policy was compatible with Peking's.

For at least twelve hours after President Kennedy's June 10 commencement talk at American University, no word appeared in the Moscow press or radio about the speech. Neither did the text of the Chinese open letter to the Russian people, which, in context, seemed to have been written in the certainty that the Russian leaders would have no choice except to make it public as part of the bitter acknowledgment of the failure of the Khrushchev policy.

The stage was thus set for one of history's dramatic turning points. During that twelve-hour period when nothing was said in the Soviet press about either the Kennedy talk or the Chinese letter, it is possible that the Khrushchev government was pondering a fateful decision. If it published the Chinese letter it meant that coexistence had failed and that an ideological closing of the ranks with Peking was inevitable. If it published the Kennedy talk, it meant that Khrushchev's personal policy had prevailed and that further measures of accommodation with Washington would be pursued.

Finally, the news broke in Moscow—the news about President Kennedy's talk. The full text was published in *Izvsetia*. The Russians had always been touchy about what they considered to be American ignorance or insensitivity to their colossal loss of life and property in World War II compared to that of the United States. When, therefore, President Kennedy paid eloquent tribute to the suffering of the Russian people in the war and to their basic yearning for peace, he touched a responsive chord in the Russian soul.

He did something more: he spoke realistically about the things that were necessary and therefore possible in a joint program for peace. He offered a limited test-ban treaty.

If the President had confined himself in his talk to the test-ban treaty, it is possible the proposal would have been rejected. But by going far beyond the treaty and by defining the realistic basis for further agreements between the two countries, the President opened up a whole catalogue of reasonable expectations. The President did not minimize the ideological differences between the two countries, nor did he provide the slightest basis for supposition that the United States might be softening its stand on Berlin or any of the other basic elements in American policy. The Cuban crisis had provided all the evidence that might be needed on that score. But what the talk did do was to recognize the imperatives of an atomic age that called for a new sense of proportion and a new awareness of a common destiny.

The positive response in the Russian press to the American University talk was followed by indications that the Soviet leaders were ready to proceed with negotiations for a limited test-ban treaty. Once again Averell Harriman was sent to Moscow to complete negotiations for the treaty, where he had three separate interviews with the Chairman between July 15 and 26. The initialing of the treaty on July 25, 1963, was a fitting tribute to Harriman's major role in obtaining Russian agreement to the new terms of the treaty.

Meanwhile, however, the treaty had to pass the United States Senate. Fifty votes seemed fairly certain. Where were the other seventeen votes coming from? A major campaign would have to be mounted. In the forefront of the President's mind was the scalding awareness of Woodrow Wilson's failure to obtain the kind of public backing that would have made likely Senate ratification of U.S. membership in the League of Nations.

Where did the public stand? Many members of Congress had the idea that the public was wary of any test ban. The heavily muscled campaign for unlimited nuclear testing over the previous six years had persuaded many Americans that any limitation on the nuclear capacity of the United States would be contrary to the requirements of the nation's military security.

Even the fact that the proposed treaty excluded underground testing failed to satisfy diehard antagonists of a test ban. A number of military spokesmen and scientists, led by Dr. Edward A. Teller, took the position that any limitation on the weapon-making potential of the United States was prejudicial to the national security.

Private polling information, however, indicated that the trend of public opinion was moving toward support of a test ban. The reasons ran all the way from apprehension over radioactive contamination to fear of a widening nuclear arms race. But this had not yet been reflected in the signs by which members of Congress judge the mood of their constituencies; i.e., mail, telephone calls, visits, or other forms of direct or indirect communication. Quite the contrary: the mailbag and visitors' tally sheet of most congressmen showed that opposition to any test ban had been about ten to one; during some weeks, as high as twenty to one.

In Washington I discussed with Pierre Salinger the problem of test-ban ratifications. We reviewed the work of the Ad Hoc Committee for a Nuclear Test Ban. He suggested that it might now be reconstituted as a citizens' committee for ratification. Salinger emphasized that not all the required Senate votes were in hand. The President could be resoundingly defeated if even a few of the many borderline senators were to go the wrong way. A whirlwind campaign for educating and mobilizing public opinion was needed. Salinger felt that the key members of the Ad Hoc Committee, together with prominent business leaders who strongly favored Senate ratification, should be brought to the White House for a meeting with the President. The purpose would be to form a citizens' committee for test-ban ratification and to decide on general strategy.

It was against this background that a small group met with the President in the Cabinet Room August 7, 1963. On the government side, in addition to the President, were McGeorge Bundy, Lawrence F. O'Brien, Frederick G. Dutton of the State Department, and Adrian S. Fisher of the Arms Control and Disarmament Agency. For the private sector, in addition to James J. Wadsworth, former U.S. Ambassador to the UN and former chief representative in test-ban negotiations, were Walter Reuther, president of the United Automobile Workers of America; William L. Clayton, former Undersecretary of State of Economic Affairs under President Truman; Marion B. Folsom, former secretary under President Eisenhower of the Department of Health, Education, and Welfare; and myself.

The President began the meeting by saying that he placed the highest importance on the fight now just getting under way for Senate ratification. Even since Woodrow Wilson, he said, a President had had to be cautious about bringing a treaty before the Senate unless he had a fairly good idea where the votes would come from. To get two-thirds of the Senate behind any issue was a difficult and dubious undertaking; to get it on a controversial treaty was almost in the nature of a miracle. He said he could name fifteen senators who would probably vote against anything linked to President Kennedy's name —"and not all of them are Republicans."

On this particular question, he said, we had to face the

fact that if a vote were held right then, it would fall far short of the necessary two-thirds. He reminded us that most senators had not made public statements in favor of a test ban. Meanwhile, most of the mail to the Congress, like the mail to the White House, was against any ban on testing. He then turned to Larry O'Brien, White House liaison aide with the Congress, and asked for a tally on congressional mailing on this issue. O'Brien said the latest figures showed the mail was approximately fifteen to one against a ban on testing.

Then the President rang for Mrs. Lincoln and asked for a tabulation of current White House mail. Mrs. Lincoln returned shortly. The President scanned the list, then looked up and smiled.

"The category that leads the list again this week is requests to the Kennedy family for money. We seem to have landed on a number of prime lists of good prospects for a touch. I also see that we have received more letters on the White House animal pets than on the financial crisis of the United Nations. Nuclear testing is far down the list. But most of the people who write on this subject are against the ban.

"My guess is that the mail to the White House and to the Congress against any nuclear test ban will increase. The opponents to any ban are going to try to snow us. I'm not going to underestimate the size of the opposition or the wallop they could pack. They know that all they need to knock out the treaty is a handful of votes.

"So I thank you for taking on a very tough job. You must not hesitate to use me in any situation where you think I can help. If there's any person or organization you want me to communicate with personally, I'll do it. I want you to stay in close touch with me. I'd like to receive regular reports on your efforts and to know of problems as they develop. Now, I'd like to hear about your plan of action."

Speaking for the group, I reviewed for the President the basic facts about the Ad Hoc Committee for a Nuclear Test Ban, which had just been reconstituted and enlarged for the purpose of mounting a national campaign for Senate ratification. I handed the President a sheet containing the names of active members of the committee, including William Bernbach, advertising agency executive; Harry Culbreth, Nationwide Insurance Company; Clark Eichelberger, executive director of the American Association for the United Nations; Dr. Maurice N. Eisendrath, president of Union of American Hebrew Congregations; David Finn, of Ruder and Finn; Dr. David R. Inglis, Argonne National Laboratories; Dr. Homer Jack, director of National Committee for a Sane Nuclear Policy; Mrs. Lenore Marshall, poet and novelist; Kenneth Maxwell, National Council of Churches; Seymour Melman, Columbia University; Mrs. Josephine W. Pomerance, A.A.U.N.; Dr. Eugene Rabinowitch, editor of *Bulletin of Atomic Scientists;* Walter Reuther, United Automobile Workers; Robert Stein, editor of *Redbook* magazine; Herman W. Steinkraus, former president of the U.S. Chamber of Commerce and then president of the A.A.U.N.; John Sullivan, New York State Commission for Human Rights; Harold Taylor, educator; Ambassador Wadsworth; Paul Walter, president of the United World Federalists; and John F. Wharton, lawyer.

The new Citizens Committee, I continued, was an enlargement of this group. It was organized on several levels. Altogether, the new committee consisted of forty-eight prominent Americans in business, education, health, the arts, science, labor, religion, and diplomacy. On it were two former cabinet members, a former chairman of the Board of Governors of the Federal Reserve System, a former undersecretary of state, a former U.S. senator, the heads of five religious denominations and the presidents of several national community organizations. Walter Reuther was working with us to get the message across to organized labor. James G. Patton was doing the same for the National Farmers Union. James B. Carey was initiating a program for the Industrial Union Department of the AFL-CIO. Will Clayton and Marion Folsom were organizing a committee of business leaders, with the help of Oscar DeLima, chairman of the board of Roger Smith Hotels, and Sol Linowitz, chairman of the board of Xerox Corporation. We were preparing material for debates, whether at meetings of the Chamber of Commerce, Lions Club or Rotary International, League of Women Voters, or various other civic groups or public forums.

Then there was the comprehensive plan for getting across the arguments favoring a test ban on a nationwide general basis. In preparation were full-page national advertisements, television materials for forum and discussion programs and debates, and special materials for weekly newspapers and rural journals.

The National Committee for a Sane Nuclear Policy, I said, had played a vigorous role in the campaign to date and would continue to do so. The President then asked for a full account of SANE's background.

I said that SANE had been formed in 1958 in order to dramatize the dangers of nuclear testing. It began as a coordinating committee for various groups interested in the question

—mostly world affairs and religious organizations. We were fortunate in having men such as Homer Jack and Donald Keys as full-time operating heads of the organization.

The immediate burst of national response to SANE's first public statements quickly catapulted the coordinating committee into a full-size national organization, with chapters and membership all over the country. In fact, SANE had grown so fast it was unable to keep track of all the local committees formed under its name. Clarence Pickett, executive secretary emeritus of the American Friends Service Committee, and I, as co-chairmen, had been somewhat flabbergasted to find ourselves with a national membership organization on our hands. The last thing in the world we wanted to do, in fact, was to create yet another organization in the foreign policy field. Pickett believed there was already too much overlapping among "peace" groups.

Like it or not, however, SANE had tapped an important vein in American public opinion. Effective though it was on the community level, it was not quite what was now needed in terms of a specific mechanism aimed exclusively at a nuclear test ban. This led to the creation of an Ad Hoc Citizens Committee for a Nuclear Test Ban, which in turn was now serving as the core for the formation of a Citizens Committee for a Nuclear Test Ban.

This recitation had taken perhaps three or four minutes. I was struck again by the fact that the President was a most sympathetic listener. When you spoke, he looked directly at you in a way that encouraged full exposition.

The President said he had known fragments of the SANE story and was glad to have all the pieces put together. He then asked whether we had adequate staffing to carry out the substantial undertaking represented by a national campaign.

I referred to David Finn, whose company (Ruder and Finn) was coordinating this total effort. Working full time with Mr. Finn in this program was Miss Lillie Shultz, who had had considerable experience in national campaigns calling for full-scale debate under deadline conditions. Mary K. Harvey, of the *Saturday Review* staff, who had been my right arm in this general field, would work directly with David Finn and Lillie Shultz. Kenneth Maxwell of the National Council of Churches, and Rabbi Maurice N. Eisendrath of the Union of American Hebrew Congregations, were organizing the religious community. We also expected important help from Archbishop John J. Wright of Pittsburgh, who had provided valuable counsel at the time SANE was formed.

The President asked whether Cardinal Richard Cushing of Boston was being drawn into the campaign.

Not to the best of our knowledge, we said.

"He ought to be," said the President. He then addressed himself to Fred Dutton. "Remind me to telephone him and see if we can't get him involved." Then, turning back to us, he asked, "How about your business leaders?"

Will Clayton and Marion Folsom, in that order, spoke of their belief that a substantial number of the nation's business leaders would associate themselves with the committee. They expressed confidence that the business community would mobilize enough support to counter any impression that the "business realists" felt the national security would not be served by a treaty.

The President asked about Edwin P. Neilan, president

of the U.S. Chamber of Commerce; how did he stand? Mr. Folsom reported that Will Clayton, Ambassador Wadsworth, Cousins, and himself had called on Mr. Neilan, and found him responsive but not fully committed. The President said he would be glad to telephone Mr. Neilan and invite him to the White House for a chat on the matter. We suggested it might be useful if he would do the same with Frederick R. Kappel, the chairman of the board of the American Telephone and Telegraph Company.

Ambassador Wadsworth said he expected that the principal opposition to a treaty would be represented by Dr. Edward Teller and by members of the military. He didn't underestimate the influence of this combination. They were bound to charge that the treaty would weaken the nation's defenses; that it would give all sorts of advantages to the Soviet Union, which intended to violate the treaty secretly; that talk of fallout as being dangerous was nonsense, etc., etc. Wadsworth felt that arguments such as these, repeated often enough over a short stretch, would add up to a powerful campaign.

The President said he would do what he could to hold his administration in line on the issue. He said that Secretary of Defense Robert S. McNamara would give strong, positive testimony, stressing the arguments about the danger of uncontrolled development and spread of nuclear weapons around the world.

Even so, the President said, we had to assume that influential members of the military, or even the Atomic Energy Commission, would do what they could behind the scenes to influence members of Congress or the press.

It's no secret, he added, that many were opposed to any limitation on the weapons-development and weapons-producing capacity of the United States. In fact, he said, some generals believed the only solution for any crisis situation was to start dropping the big bombs. When the President would pursue the matter by asking how bombing would solve the problem, the replies would be far less confident or articulate.

Be that as it may, our committee could not assume that the forces opposed to a test ban were not substantial, or that they would not receive powerful aid, open or otherwise, from those who believed in the unrestricted use of military power. And the President agreed with Ambassador Wadsworth that Dr. Teller would be a hard, driving, difficult force to counter.

The President then summed up. He reiterated the need for important business support and suggested a dozen names. He said that scientists such as James R. Killian, former president of Massachusetts Institute of Technology, and Professor George B. Kistiakowsky of Harvard University, would be especially effective if they could be recruited. He felt that religious figures, farmers, educators, and labor leaders all had key roles to play and he mentioned a half dozen or more names in each category. Then he went down the list of states in which he felt extra effort was required. He was glad that organizations such as SANE, the Americans for Democratic Action, the United World Federalists, the American Association for the United Nations, and the American Friends Service Committee were going to continue to give full support to the issue, but he wanted to be sure that they did not make the test ban appear to be solely a liberal cause.

We got up to leave. The President thanked us again and reemphasized his wish that we report regularly and directly to him.

Outside the Cabinet Room the press was waiting. According to custom, we did not quote the President directly, but said we were confident that our committee would receive complete backing from the White House. We also emphasized that at the heart of our campaign was the belief that the security of the United States depended more on the control of force than on the pursuit of force.

We were asked what our response was to Dr. Teller's contention that the dangers of radioactive fallout had been grossly exaggerated and that the amount of fallout from the tests so far, in its effect on the individual, was no greater than would be given off from the luminous dial of a wrist watch. We replied that fallout from nuclear weapons was not merely external, as in the case of emanations from a wrist watch, but that radioactive strontium, cesium, and iodine, to mention only a few, were infecting water, food, and milk, and had already turned up in children's bones. Moreover, the effect was cumulative. The body did not eliminate all the strontium or cesium it ingested. The poisonous strontium, for example, was mistaken by the body for calcium in the bone-building process.

For the next seven weeks, the debate over testing dominated the news. We were surprised to find that newspaper editorial support was stronger and more widespread than advance indications had led us to believe. Against this was the fact that Dr. Teller had enlisted far more allies than we had anticipated.

The new owners of the *Saturday Review*, the McCall Corporation, gave me unstinting support. Robert Stein, editor of *Redbook* magazine which, with *McCall's* and *SR*, comprised the magazine properties of the McCall Corporation, played an important role in the campaign. Bob arranged for a special meeting between the President and the editors of the nation's leading women's magazines, representing a reading audience of seventy million or more. The President explained to the editors why a test ban was urgently necessary and then answered their questions. The transcripts of the meeting appeared simultaneously in all the women's magazines. It was the first time that the influential women's magazines had come together in a joint publishing venture of this kind.

Mary Harvey, as I indicated earlier, had been my lieutenant for several years, both on the magazine and in various projects such as the Dartmouth conference series. Mary had a wide background in the fields of communications and public policy. She coordinated the work of the Ad Hoc Test Ban Committee, which led directly to the formation of the Citizens Committee for a Nuclear Test Ban Treaty. After Mary left to join *McCall's*, Dori Lewis became my liaison in the campaign for ratification. She gave me full briefings on every situation along the way, helped prepare the progress reports, and supplied essential connective tissue for the committee.

Lillie Shultz, working with Ruder and Finn, was a superb craftsman in the field of public opinion. She drew up a comprehensive blueprint for organization and action; she had a rare talent for avoiding false starts and dead ends. When she moved in a certain direction we could be sure things would happen. I had great respect for David Finn's judgement in recruiting her from outside his own company for this purpose.

Where did the money come from? Will Clayton and Marion Folsom were instrumental in raising important sums. So was Walter Reuther. I doubt, though, that anyone worked harder or more effectively in money-raising than Lenore Marshall, the poet, whose commitment to the goal of a test ban since 1958 had been a major source of impetus to the entire movement. Mrs. Josephine W. Pomerantz, as secretary to the Citizens Committee, also made generous donations.

On August 27, 1963, the leaders of the test ban campaign met in Ambassador Wadsworth's office in Washington, which had served as national headquarters for the committee, to take soundings on the existing situation and to map plans for the final weeks of the drive. On the plus side we could take some measure of satisfaction from the fact that the tide of congressional mail had at last begun to turn. In six weeks the ratio of fifteen or more to one against a cessation of nuclear testing had shrunk to about three to two against, taking the Congress as a whole. There was every reason to believe that the trend would continue to move in our favor. The public opinion polls showed a steady rise in the percentage of people now in favor of a treaty. If the trend held up, there would be a slight preponderance within a month or so.

The following progress report was sent to the President on August 28, 1963:

MEMORANDUM TO: The President

SUBJECT: Progress report on your specific suggestions for the public campaign to ratify the test-ban treaty.

I.   *You emphasized the need for support of business leaders. This is what was done:*

    A. Marion Folsom and Will Clayton sent wires and letters to the nation's top businessmen, inviting their participation in statements of support for ratification. Forty-five responded affirmatively.

B. The statement of support from business leaders was released to the press on August 14. The wire services gave adequate coverage to the story.

C. The statement from business leaders served as the basis for a double-page advertisement in the *Washington Post* and *The New York Times*.

D. The same advertisement from business support with the addition of important local names also appeared in a full single page in the *Chicago Tribune* on August 23, following your suggestion to Fred Dutton.

E. Business leaders in Chicago, St. Louis, Indianapolis, and Des Moines are organizing local campaigns. This includes public statements (to be scheduled) and direct communication with the senators.

II. *You spoke of the need to enlist the support of scientists like President Killian and Professor Kistiakowsky. This is what was done:*

A. President Killian agreed to serve as Chairman of the Scientists Committee for A Nuclear Test Ban. A number of scientists accepted, including Professor Kistiakowsky.

B. The Scientists' Committee is issuing a news release at the time of the full-page newspaper advertisement.

C. Professor Rabi of Columbia University brought together a group of Noble Laureates for a similar purpose.

III. *You suggested that religious leaders be given prominent attention, with full representation from each of the faiths. This is what was done:*

A. Dr. Kenneth Maxwell of the National Council of Churches sent out urgent action bulletins calling for sermons and extensive letter-writing to members of Congress.

B. Dr. Maxwell has been attempting to coordinate the efforts of other denominations with a view toward issuing a combined statement.

C. The Methodist Church has been coordinating a special effort to develop public support for the ratification.

D. Cardinal Ritter has discussed the matter in general terms with Archbishop Wright of Pittsburgh. Archbishop Wright, in communication with Father Morlion and myself, said he is communicating with a number of Catholic bishops who will join with other religious leaders in support of the treaty.

E. I had a good talk with Father Cronin of the National Catholic Welfare Conference. He agreed to do all the circulating work necessary to solicit important Catholic names.

F. Rabbi Eisendrath, president of the Union for American Hebrew Congregations, has called on all members for widest possible campaign of letter-writing to Congress. Rabbi Uri Miller, Synagogue Council of America, is doing the same.

IV. *You stressed the importance of reaching farmers. This is what was done:*

A. A general strategical plan was drawn up. The plan was divided into two general parts—one concerned with information and education, the other with a program for political action.

B. A special news and information kit was prepared for use for farm newspapers, daily and weekly and for radio and television farm programs.

C. Radio tapes featuring statements by the President, Secretary Rusk, Governor Harriman, William Foster, and senators from the areas.

D. Angus McDonald sent out action bulletins to members of the Farmers Union calling for letters and visits to senators.

V. *You suggested that scholars and university presidents be enlisted. This is what was done:*

A. President Goheen of Princeton agreed to serve as chairman of the University Committee for a nuclear test ban. Sixty-three names are on his prime list.

B. The educators' statement will be given prominent attention through news release and advertisement.

VI. *You suggested that the Machinists' Union might be involved and that the St. Louis Post-Dispatch might be effective. This is what was done:*

A. Walter Reuther communicated with the Machinists' Union which made known its position to Senator Jackson.

B. The *St. Louis Post-Dispatch* has published editorials against the opponents of the limited test-ban.

C. A group of St. Louis businessmen have come together under the leadership of Raymond Wittcoff, president of Transurban Development, Inc. Several members of this group have communicated with senators personally. Mr. Wittcoff has arranged for the publication of the business leaders' advertisement in the local press.

VII. *You asked for pinpointed action on the key states. You suggested a wide range of activities directed to winning over the wavering senators and heating up the lukewarm ones. This is what was done:*

Beginning on August 9, the efforts of a special task force were devoted to securing support for the treaty in critical states, among them: Colorado, Illinois, Iowa, Missouri, Ohio, South Dakota, and Washington.

The following was done:

A. Survey of the area.
   1. Politically, the area was largely Republican.
   2. An examination of the economic structure indicated that the population would be influenced largely by farm and related organizations, labor unions, as well as the press.

B. Solicitation of the cooperation of the Department of Agriculture.
   1. Through the good offices of Secretary Orville Freeman, his personal assistant George Barnes called together the scientists in the department to prepare a memorandum on the need for putting an end to radioactive fallout. This was used as the basis for a feature article which subsequently appeared in farm weeklies.

C. Approaches to farm organizations and related agencies.
   1. A kit of relevant materials was provided the

Farmers Union organizations in the seven states for use in its press.

2. James Patton, president of the union, was thoroughly briefed on the problem and encouraged to make personal efforts, particularly in Colorado where his influence was considerable, both with the two senators and with Palmer Hoyt, editor of the *Denver Post*.

3. A meeting was held with a top executive of the Pure Milk Association in Chicago. Subsequently, using the committee's material, this executive got in touch with various individuals and organizations urging them to influence their senators. The concentration was on Illinois, Colorado, and Washington. In the opinion of the executive, this activity helped produce the affirmative votes of Senators Dirksen, Jackson, Dominick, and Allot.

D. Provision of broadcast and print material for use by the press.

1. 2,187 informational kits were distributed as follows:

    (a) To 1,266 weeklies with a circulation of 1,000 and over.

    (b) To 315 dailies.

    (c) To 154 farm editors of dailies and farm publications.

    (d) To 371 radio and TV stations.

    (e) To 81 radio and TV farm programs.

VIII. *You suggested that liberal organizations such as SANE, UWF, and ADA be utilized expeditiously and carefully. This is what was done:*

A. The liberal organizations, including representatives of labor unions, SANE, UWF, ADA, American Association for the United Nations, Friends Committee on National Legislation, among others, have come together in a working committee in Washington. This committee fully accepts the need to work behind the scenes, activating their memberships, and cooperating in a combined program to produce support for the treaty. The organizations have given priority to the writing of letters, organizing home letter-writing, meetings with senators, and newspapers the main targets.

B. Correspondence with Senator Jackson.

C. Dr. Spock's report on Physicians for Social Responsibility.

D. Paul Newman's letter sent to show business celebrities.

E. Report of the Washington lobby.

F. Spot ads and editorial mats available to local media.

*   *   *

Each day's progress reports brought encouraging news, but there was danger of overconfidence. For if the fact that the battle was moving our way was known to us, it was certainly known to the other side. Prominent newspaper stories quoted unnamed military sources who declared that our defense program was tied to new developments in nuclear warfare, for which nuclear testing was indispensable.

Dr. Teller and I had several public confrontations on television and elsewhere. I fear our audience found him the more restrained of the two. There came a point in our meetings when I was unable to resist full and unambiguous response. Teller would use his authority as an atomic scientist—invariably he would be billed as the "father of the hydrogen bomb"—to reassure people about radioactive fallout and then go on to tell of the peril to the nation's security if we should give the Soviet Union the military advantage of a test ban which it could probably circumvent.

Fortunately, most of Dr. Teller's colleagues in the fraternity of nuclear scientists disagreed with him. A special committee of scientists attached to our group took up Teller's technical arguments one by one. We gave wide distribution, for example, to the views of such scientists as Dr. Harrison Brown of the California Institute of Technology; Dr. Leo Szilard, one of the pioneers in the development of nuclear energy; Dr. Eugene Rabinovitch of the Department of Bio-Chemistry, University of Illinois, and editor of *The Bulletin of the Atomic Scientists;* Dr. Hugh C. Wolfe, chairman of the Department of Physics at the Cooper Union School of Engineering; Dr. Barry Commoner of George Washington University, St. Louis; and Dr. David R. Inglis of the Argonne National Laboratories.

Meanwhile, Dr. Linus C. Pauling, Nobel Prize-winning scientist, was making strategic use of his worldwide petition, presented to the UN Secretary-General in 1958, signed by nine thousand leading scientists from more than forty countries, and demanding an international agreement to stop nuclear tests.

Dr. Teller was valiant and energetic—I could give him credit for that—but his effort to clinch the scientific argument for continued testing was running into trouble.

On a broader front, what we were up against was the

problem identified by President Kennedy in his discussion with our committee at the start of the campaign. That question was whether any limits were to be placed on the development of American weaponry. What we were really debating was not whether nuclear testing produced dangerous radioactive fallout; nor whether it would be possible for the Soviet Union to undertake atmosphere tests secretly; nor whether the continuation of testing made it impossible to embark on further measures to halt the spread of nuclear weapons around the world—but whether the life of the nation depended on the control of force or the unimpeded pursuit of it.

The campaign culminated in the hearings before the Senate Foreign Relations Committee. The two principal witnesses against ratification were Dr. Edward Teller and General Thomas S. Power, Strategic Air Command Chief. General Curtis E. LeMay, Air Force Chief of Staff, acknowledged opposition to the treaty, but said he had no desire to pursue the matter since the document had already been initialed.

Supporting ratification was a wide array of government witnesses, led by Secretary of State Dean Rusk, Secretary of Defense Robert S. McNamara, and Chairman of the Joint Chiefs of Staff General Maxwell D. Taylor. For weeks some high-ranking members of the military had been quoted directly or indirectly—on the floor of the Congress and in the press—about the serious threat to the national security represented by a test ban. Secretary McNamara refuted such statements; his testimony was probably the most important single factor leading to the favorable report on the proposed treaty by the Senate Committee. Both former Presidents Harry S. Truman and Dwight D. Eisenhower wrote letters to President Kennedy and Senator J. W. Fulbright, respectively, supporting ratification.

Before testifying in behalf of our citizens' group, I spoke to the President on the telephone. He urged me to connect the test-ban issue to the larger question of East-West relations, as I had done in talking to him shortly after my return from the second visit to Khrushchev.

My testimony before the Senate Foreign Relations Committee followed the general lines of the President's suggestion. Most of the questioning by the committee was directed to the Soviet side of the test ban. Why did the Soviet Union want the pact in the first place? What were the implications of the Soviet-Chinese rift as they concerned or affected a treaty to outlaw nuclear testing? Were the differences between Stalin and Khrushchev as substantial as they appeared?

I made it clear that while my answers to most of these questions were speculative rather than definitive, there were substantial indications that it was as much in the national interest of the Soviet Union to have a nuclear test ban as it was in the American interest.

Several days before the treaty came to a vote in the Senate, it appeared probable that sufficient votes for ratification were assured. Even at this point, however, the President urged us to take nothing for granted and to keep on working until the last moment.

Ratification came on September 24, 1963, by eighty to nineteen, a substantial margin over the required number of votes. There were some surprises. Senator Margaret Chase Smith (R., Maine), whose views on foreign policy were progressive, and Senator Frank Lausche (D., Ohio) voted against ratification. Among those voting for it, however, were Senators Everett Dirksen (R., Illinois) and Karl E. Mundt (R., S. Dak.).

The positive significance of the victorious fight for a nuclear test ban was represented not just by the prospects of decreased atmospheric contamination but by the encouragement it provided that even more fundamental elements involved in the making of peace, as the President saw it, were now within range. Effective agreements on workable disarmament; on measures to stop the spread of nuclear weapons; on East-West trade; on enlarged cultural exchange; on new approaches to a settlement in Indo-China; on the Berlin question; on mutual development of space for peaceful purposes; on increased respect for and use of the machinery of the United Nations—all these now seemed ready for fruitful negotiation.

During the summer and early fall of 1963 President Kennedy carefully laid the groundwork for pursuing initiatives in all these directions. Those who worked closely with him reported he didn't minimize the difficulties or the complications, but neither did he undervalue or underestimate the size and reality of the hopeful prospects that seemed to open out before the United States and the world. This optimism, in fact, was reflected in his last public speech in Dallas, just before the assassination.

*W*HAT CONCLUSIONS can be drawn from the interaction of President Kennedy, Premier Khrushchev, and Pope John, and the consequent upturn in the prospects for world peace in the year following the Cuban crisis, of which the limited ban on nuclear testing was perhaps the most significant achievement?

What seems most striking about the episode described in this book is not just the rapidity with which international tensions were eased once world leaders energetically moved in that direction but the fact that Khrushchev and Kennedy went against powerful cross-currents within their own countries to do so. Pope John XXIII, too, broke with orthodoxy in an effort to reduce the danger of war.

It is obvious, of course, that all the world's political leaders have potentially powerful moral initiatives within their reach. It is equally obvious that, once they decide to use these initiatives for a common purpose, historic changes can be brought about. What is less obvious is the fact of multiple restraints and obstructions to the exercise of such initiatives. The natural tendency of a national sovereignty is to assert its will regardless of abstract definitions of right or wrong. Nothing is more characteristic of a sovereign state than its habit of defining morality as that which serves the national interest. When the Soviet Union moved into Czechoslovakia with military force to suppress the desire of the Czech people for political and cultural freedom, the Soviet leaders did what they thought was necessary for their political and military security; anyone in the inner councils who raised the question of morality would no doubt have seemed fatuous. When Italy rained bombs on Ethiopian villages, the only question was whether the bombing would accomplish Italy's purpose—which was to terrorize the Ethiopians into submission. When the United States embarked on its air bombing in both South and North Vietnam, the decision had its source in U.S. military-political objectives. The moral question about the bombing, however, will continue to be raised by history.

What I am suggesting is that the making of moral judgments is not a natural function of sovereign national states. Yet progress and indeed survival depend ultimately on the ability of society to make moral judgments. The same society or government that has developed abstract ideas of justice within its own jurisdiction does not accept moral law as a primary factor or obligation as a governing principle in its world policy. The reasons are not obscure. First, the state recognizes that the actions of other nations also originate in self-interest and not in morality. Second, the sovereign nation acts out of a calculus of power that tends to obliterate moral imperatives. When Premier Khrushchev, in the Gagra sessions described earlier in this book, said that his generals and his scientists were pressing him to continue tests in the interests of Soviet security, he happened also to describe an identical situation in the United States. The job of the generals and military scientists—in whatever country—is to proceed on the assumption that the security of the nation rests on their access to adequate weaponry and manpower. There is inevitable resistance to any measures that seek to curtail or limit such power.

Military power becomes political power not solely because of maneuvering inside government but because the society itself tends to metabolize military power on the economic level. The biggest impetus for large military spending programs in the U.S. has come not just from industrial military contractors but from men and women who feared they would lose their jobs.

Coinciding with this institutionalization of power is the ease with which national moods and tempers can be exacerbated and inflamed. Any hostile statement by high officials of one nation against another is certain to be magnified in the nation thus abused. Such statements are generally regarded as proof of the aggressive designs by the first nation against the second—designs and threats which are usually reciprocated by the second nation. This mutual suspicion and hostility are escalated by those on both sides whose positions and policies, and thus their power, are strengthened and ostensibly confirmed by the resultant insecurity and tensions.

The Cuban missile crisis of 1962 was a culmination of this process. Whether or not its specific cause was the attempt by the Soviet Union to forestall an American invasion of Cuba, as Premier Khrushchev had said; or an attempt to confront the United States with the same kind of nearby missile threat that the U.S. maintained in Turkey for possible use against the U.S.S.R.; or a probe by the Soviet Union, as some observers have suggested, to test American nerve in a serious challenge to its historic hemispheric position—whatever the specific cause or causes, the Cuban crisis demonstrated the ease with which the world could slide into a nuclear crisis. It also demonstrated the vulnerability of mankind to clashes of national interests.

What was most terrifying about the Cuban crisis, therefore, is that there was nothing unnatural about it. Both countries were proceeding according to the historic position by which nations assert or maintain their security requirements or national ambitions. President Kennedy and the group around him were not unmindful of the possible chain of events that might lead to a nuclear holocaust. Night after night, he met with his staff and closest advisors in an attempt to find a formula that might avert the use of thermonuclear force. The same nightly conferences were held in the Kremlin. Also, as we have seen from an earlier chapter, Khrushchev had to go against the advice of top military and party officials in writing the letter to Kennedy that resulted in withdrawal of the missiles. It is significant, too, that in the months following the Cuban crisis, Khrushchev was under heavy criticism from the Chinese leaders inside the Communist world for appearing to lack courage when confronted by the "U.S. paper tiger." It was the kind of taunt that was certain to hit a sensitive nerve, for the last thing in the world any head of state wishes to have said about him is that he sacrificed the national interest rather than face up to a dangerous challenge. Most of Chairman Khrushchev's talk to the Supreme Soviet on December 12, 1962, was in the nature of a refutation of the Chinese accusation that he had shown the white feather.

The fact that a combination of adolescent taunting and dangerous irrationality should figure in the behavior of nations may seem ludicrous; but nations involved in confrontation situations exhibit all the petulance, arbitrariness, irrationality, and false pride associated with the immature mind. What adds to the tragedy is that aggregations of people inside nations become caught up in such irresponsible emotional outbursts. No ego is more powerful than the group ego.

What makes the personal exchanges between Kennedy and Khrushchev so remarkable, therefore, is that both men had to overcome severe opposition within their own establishments in order to reduce the hostilities and tensions between the two societies. They had to contend with all those, in and out of government, whose careers and not just their convictions might be upset by an official policy of friendly relations.

Even Pope John had to break with tradition in order to attempt to intervene directly in the Cuban crisis. More unorthodox still was the mission to Moscow he sponsored for the purpose of establishing more than superficial contacts with a Communist world power. In a sense, Pope John's attempt to open up lines of communication with the Kremlin was a reflection of the ecumenical revolution associated with his name. But the exchange between Pope John and Khrushchev was somewhat startling even in the context of the Pope's determination to throw open the windows of the Church and Khrushchev's determination to blast open the closed policy of Josef Stalin.

What emerges most of all from the experiences of 1962–63 is the fact that the future of mankind is too important to be left to the kind of historical accident that brought three men together who could transcend the limitations and obstructions imposed by tradition and established organization. What is necessary today is not the de-institutionalization of the nation but the institutionalization of humankind. The sovereign state does not have to be abolished in order that the human interest on earth should come before the national interest. But so long as the sovereign state represents the ultimate form of human organization, the condition of the human habitat will be in jeopardy. The control of power rather than the pursuit of power now becomes the first order of international business.

The United Nations, in the minds of the world's peoples, if not their statesmen, was created for such a purpose. The full development of the United Nations, therefore, into a responsible and effective organization is the essential condition for a world that has to be freed from the terror of international breakdown in a nuclear age. No one can minimize the difficulties of transforming the United Nations from an organization that reflects the power struggle among nations to an organization that is in a position to moderate or resolve that struggle; but neither can anyone minimize the difficulties of safeguarding the human estate without it.

The ultimate lesson of Kennedy, Khrushchev, and Pope John, therefore, is that the world must not assume that men like them will automatically spring into action at a point of maximum danger. Reason demands that proper instruments be devised for dealing with basic causes of war and for creating the new institutions that can serve humankind as a whole.

Some Notes on the Background of the Campaign for the Nuclear Test-Ban Treaty

*I*T IS HARD TO FIX a precise date for the effective beginning of the campaign to put an end to the test explosions of nuclear bombs. One reason, perhaps, is that there was no clear-cut distinction between the effort to outlaw nuclear weapons altogether, dating from the explosion over Hiroshima, and the more particularized campaign against testing.

The effective beginning of public awareness of the fact of radioactive fallout, however, probably dates from the explosion of a hydrogen bomb in the Bikini test area in the Pacific Ocean on March 1, 1954. One result of that explosion was that twenty-three Japanese fishermen, aboard the tuna trawler *Fukuryu Maru* (*Lucky Dragon*) were showered by hot radioactive ashes. Another result was that Marshall Islanders were hit by the fallout, even though they had been moved to a supposedly safe site.

The news about the Japanese fishermen came first. The men had severe symptoms of the world's newest and potentially most lethal disease—radiation sickness. The symptoms included fever, presistent vomiting, hemorrhaging, debility, exfoliation, infection, and alterations in the composition of the blood.

It was said that the *Lucky Dragon* had ventured far into the prohibited zone, but this turned out to be incorrect. Appropriate expressions of dismay and regret over the incident were offered by the U.S. government, but the world was reminded that the tests were essential to American military security. One thing, however, was clear: Whatever military information may or may not have been yielded by the bomb tests, it was now proved that goverment statements concerning margins of safety for radioactivity were wide of the mark to the point of being irresponsible.

The Marshall Islanders, wards in effect of the United States government under a United Nations trusteeship, lived in an area far enough removed from the testing area to be beyond danger. But perverse winds carried the poisons over Rongelap Island and the people were hit just the same, with results not dissimilar from those that affected the Japanese fishermen. It was eleven months, incidentally, before the United States government officially admitted that the test explosions had produced unexpected fallout.

In July, 1946, the first in a series of United States test explosions was held in the Bikini area. The correspondents' ship to which I was assigned, the U.S.S. *Appalachian*, was stationed only fourteen miles from the nuclear drop. The day after the first test explosion, we cruised into the Bikini lagoon. The men went ashore. James V. Forrestal, Secretary of the Navy, said it might be nice to go for a swim. Once he dived in, countless other members of the party followed. These were the very waters which only twenty-four hours earlier had been the site of an atomic explosion. Yet so little importance up to that time had been given to the radiological power of nuclear bombs, as distinguished from the blast effects, that a member of the President's cabinet and perhaps thirty men exposed themselves to what could have been serious consequences. A simple Geiger counter reading, taken in *one* place, was apparently considered an adequate safeguard at that time. A few years later such behavior would have been considered unthinkable.

One result of the *Lucky Dragon* and Marshall Islands tragedies was that they led to a mounting demand in world public opinion for fresh efforts to bring nuclear weapons under control. President Dwight D. Eisenhower appointed Harold E. Stassen, former governor of Minnesota and presidential aspirant, as his special assistant for pursuing peace initiatives, especially in disarmament matters. After the death of Joseph Stalin in 1953, such initiatives seemed more propitious than at any time since the end of the war. It seemed apparent that the new Soviet leaders realized they could not meet production goals in industry and agriculture except by cutting back on the armaments program.

High-level discussions, however, were deferred for one reason or another for almost three years. Finally, discreet probes indicated the feasibility of direct negotiations among the nuclear powers. Negotiations of the United Nations Disarmament Commission Subcommittee were held in London from March 19 to May 4, 1956, then were resumed March 18, 1957, and continued to August 27, 1957. Represented at the meetings, in addition to the United States and the Soviet Union, were the United Kingdom, France, and Canada.

After many weeks of ostensibly fruitless negotiations in London, the Soviet Union suddenly agreed to the proposal for a test ban. Governor Stassen felt a major breakthrough had occurred. Interviewed by reporters, he reflected reasonable optimism. The resultant headlines in Washington, however, must have created some concern in the United States Government, for Governor Stassen received new instructions to ask for a "package deal" going beyond a treaty on nuclear testing and seeking comprehensive agreements on a series of items, including a ban on the production of fissionable materials and a cutback in existing weapons. Observers at the London conference were mystified. I happened to be in London at the time and went to see Governor Stassen.

"We can have an agreement to ban nuclear tests," he said, "but my instructions are to go for the whole ball of wax."

"Wouldn't there be an advantage," I asked, "in getting a test ban now that the Russians are willing to agree, then going on to other things involved in the arms race, point by point? Isn't it going to be practically impossible to negotiate the whole list at one time?"

"We'll do our best," he said.

I sought out Stephen G. Benedict, the Governor's assistant. I had known Benedict in the early days of the United World Federalists, when he was one of the student leaders. He was intelligent, resourceful, constructive.

"I'm a little puzzled, Steve," I said. "If we can get the Russians to agree to a ban on testing now, why don't we grab it?"

"We're gambling they'll agree to all the other things on the list, too," he said.

"Maybe they will, but in that case it makes even more sense to go at it point by point. How can you negotiate any other way?"

Steve said nothing for a moment or two.

"The American position is that we think it's worth asking for agreement on the complete package," he said finally. "At least it's worth the gamble."

"But it's almost certain that we'll get nothing. I can't imagine our own negotiators agreeing if the Russians had asked for a package deal. In the first place, a ban on the production of fissionable materials takes in so much ground that the negotiators could get stuck on that point alone. Does it mean a ban on the produciton of *all* fissionable materials, including those tied to peaceful purposes? On the matter of cutbacks, how will the stockpiles be monitored? At least with nuclear testing we've got something that is virtually self-enforcing. Isn't this the only place to take hold if we want to head off the spread of nuclear weapons?"

Steve looked at me and shrugged.

A few days later the meetings in London broke off. It would be a mistake, however, to regard the London negotiations as a complete failure. Governor Stassen was one of the first American negotiators to deal with the Russians outside the context of the cold war; his personal relationships with the Russians were excellent. Not infrequently he was able, away from the negotiating table, to explain a point in a way that led to progress at the regular sessions. In any case, the London meetings succeeded in laying basic groundwork that was useful during the negotiations for a limited test-ban treaty several years later.

I have no way of knowing why the United States backed away from an agreement at London. Stories in the press said that Secretary Dulles never really wanted a treaty, disagreeing with the President, but that he was willing to have Governor Stassen negotiate with the Russians, because of the requirements of world public opinion, so long as nothing of consequence came out of the negotiations. Other speculations referred to the possibility that the negotiations had antagonized Great Britain and France because only the U.S. and U.S.S.R. were involved.

Whatever the historical verdict on the London conference, President's Eisenhower's hopes for early agreement with the Russians in the field of arms control went unrealized. In his second inaugural address the President had called for a relaxation of tensions and for effective disarmament. Moreover, he later again appointed Governor Stassen to do the very job that Secretary Dulles had been reported as having prevented him from doing at Paris.

It was not until after Secretary Dulles's death in May 1959 that the Eisenhower style began to come through in the nation's foreign policy. The President became his own Secretary of State, in effect, and began to probe for effective openings. I saw the President in September 1959 not long after the death of Mr. Dulles. The purpose of my visit to the White House was to report on a trip to the Soviet Union, from which I had just returned on the official cultural-exchange program, then still in its early stages. Never had I seen the President in such fine form. He felt that recent changes in the Soviet Union presented the United States with important openings and opportunities. The new Soviet leaders, he believed, were less interested in Marxist dogma than they were in raising the living standards of the Russian people. The only way this could be done would be by cutting back on armaments. It was in the American interest, therefore, to improve relations with the Soviet Union, for it would have the effect of encouraging the

Soviets to reduce their vast military drive and the consequent danger to the United States and the world.

As for the Chinese, President Eisenhower said he agreed with that section of my written report to him in which I spoke of the rapidly developing ideological differences between the two major Communist nations. He felt that everything must be done to strengthen the Russian leaders in resisting the demands of the Chinese to return to the orthodox Communist position in favor of world revolution.

When I remarked to the President that he seemed in excellent health and spirits, he said he felt his once-serious illnesses were well behind him. In fact, he thought his heart attack of 1955 was induced less by overwork or strain than by frustration and suppressed rage. He traced the events leading to his collapse. The world in the White House had been unusually busy for several weeks, he said. He had hoped to be able to get away for a few days' relaxation and exercise, but each time he thought he saw a chance to get away, something intervened. Finally there was a break in the calendar and the President took off for Denver, Colorado, for a short holiday. Before leaving he told his aides that he hoped he would not be interrupted for routine matters.

On the first morning, the President got out on the golf course and exulted in the clear, brisk air and the opportunity to get some exercise. It took three holes to limber up. On the fourth hole, just as he was getting ready to tee off, a messenger in a golf cart rushed up to say that an urgent telephone call for the President had just come through at the clubhouse.

The President drove back to the clubhouse, speculating on the nature of the emergency. When he picked up the telephone no one was at the other end. He put through a call to the White House but couldn't find anyone who knew about an emergency call.

The President returned to the fourth tee and proceeded to dub his shot. Two holes later a courier arrived again with word that the important telephone call was now ready.

Once again the President rushed to the clubhouse. This time it was an aide at the State Department. He said that Secretary Dulles had placed the call originally but could not now be located. The aide said that so far as he knew the purpose of the call was to inform the President about a request made by an ambassador for a meeting with the President.

"Is this in the nature of a crisis or an emergency?"

"I don't think so," the aide replied.

"Will it hold until I speak to the Secretary?" the President asked.

"Yes, sir," the aide replied.

The President tried somewhat unsuccessfully to smother his rage and rang off. No sooner had he done so than he was furious at himself for having lost his temper. The aide was not responsible, he realized. That night the President had a heart attack. He didn't pretend to be an expert in these matters, but he felt his rage may have had something to do with it. The cardiac specialists were not inclined to disagree.

At any rate, there was no frustration or uncertainty in the President's manner now. He gave every impression of a man who enjoyed his job, was in full command, and was now in a position to put his own ideas and policies into effect. Confidence and optimism were in his voice and manner as he spoke about his conviction that the time might now be right for new measures aimed at an easing of tensions.

In the next few weeks the President took initiatives looking to an exchange of visits with Nikita Khrushchev. The Khrushchev visit to the United States in September 1959 confirmed the President's view that important agreements could be effected between the Western nations and the Soviet Union. Such agreements might actually lay the basis for a more durable peace. This design, however, never materialized. On the eve of the Big Four Conference in Paris in May, 1960, an American plane, especially equipped for taking photographs of strategic installations from great heights, penetrated the air space of the Soviet Union and was shot down.

The heads of the nations, including Premier Khrushchev, assembled in Paris. Khrushchev, however, refused to begin the sessions until President Eisenhower personally apologized for the incident or assured the Chairman he had nothing to do with it and that it wouldn't happen again. However, Allen Dulles, director of the Central Intelligence Agency, under whose orders the U-2 plane had operated, persuaded President Eisenhower to accept public responsibility for the incident. The argument was that the prestige and the authority of the Presidency would be downgraded if it were acknowledged that President Eisenhower was uninformed about important actions carried out by the United States government. Actually, the President had known about the U-2 flights when they began some months earlier. What he hadn't known was that they were still being carried on at precisely the time he was seeking basic agreements with the Soviet Union. There is no doubt it would have been embarrassing to have admitted he was unaware of the particular incident. On the scales of history, however, such embarrassment might be considered inconsequential alongside the significance of the collapse of the Paris Conference. For, as soon as the President accepted responsibility for the U-2 incident, Khrushchev refused to proceed with the meeting. He had assured the Central Committee of the Communist Party that the President had had nothing to do with the incident. Under the circumstances of the President's own statement, however, Khrushchev felt that there was no longer a basis for friendly discussions.

The failure of the Paris Conference was all the more regrettable because of the forward thrust that had been developed as the result not only of the President's personal diplomacy but of the United Nations' report on effects of fallout; the positive report of a joint United States-U.S.S.R. scientific team which declared a test ban technically feasible; the existence of a growing world public opinion; and the long series of meetings at Geneva. Between October 31, 1958, and September 9, 1961, some 340 negotiating sessions were held at Geneva under the auspices of the Conference on the Discontinuance of Nuclear Weapons Tests. Participating nations were the United States, the Soviet Union, and the United Kingdom. The meetings covered problems of international inspection and control, scientific detection of underground tests, techniques of on-site inspection, methods for ascertaining whether tremors were caused by earthquakes or nuclear explosives, high-altitude explosions, how and when "control posts" would be set up, etc.

A substantial measure of agreement was reached during the long period of Geneva negotiations on all points except the key one; namely, inspection and control. The Russians charged that the United States wanted inspection without disarmament, and the United States charged that the Soviet Union wanted disarmament without inspection. The Soviet representatives claimed that the United States was using the inspection demand as a subterfuge for its real objective—to get the kind of access to the Soviet Union it couldn't get otherwise. The United States replied that any realistic plan for arms control had to begin with a responsible plan for inspection.

Behind the impasse over inspections was the fact that each year hundreds of natural earth tremors occurred throughout the wide expanse of the Soviet Union. These tremors were strong enough to be picked up on seismic instruments far outside the Soviet Union. The United States position was that on-the-spot investigations were essential to rule out the possibility that some of such tremors were produced by underground nuclear explosions.

The Soviet position was that the United States argument represented an assumption of bad faith, and that it was far more logical to suppose that nations entering into an agreement in their rational self-interest would be likely to honor it.

Ambassador James J. Wadsworth, who represented the United States at various disarmament conferences from 1954 to 1960, told the Geneva conference that fear and suspicion were facts of international life and that, indeed, if tensions did not exist, nuclear weapons would not have been produced in the first place.

After much discussion the principle of on-site inspection was accepted at Geneva. The big question, however, was: just how much inspection was required as an effective deterrent against secret violations? Assuming that one hundred or more earth tremors might occur in a large land mass such as the Soviet Union, all parties agreed that it would be unreasonable and, in any case, impossible and unworkable to send an inspection team into the Soviet Union for the purpose of checking every one of these tremors. But a token number of allowable inspections—perhaps no more than a fraction of the tremors—would be sufficient to discourage clandestine test explosions, especially since the country engaging in violations would not know where or when the inspections would take place.

How many inspections? The United States had to be satisfied that this number would be large enough to discourage the Soviet Union from sneak testing, and the Soviet Union had to be satisfied that the United States was not using inspection as a pretext for espionage or general surveillance.

The United States said a quota of twenty inspections was a reasonable number. The Soviet Union thought that three were more than enough. On this point the Geneva negotiations were deadlocked month after month. All other points involved in a treaty appeared to be within negotiating range. On the question of inspections, however, the conference became stalled.

Throughout most of the Geneva negotiations the United States had the advantage of having as its representative a man who knew the importance of maintaining a cordial atmosphere and who was able to convince the men at the other side of the table that he had confidence that the differences could eventually be resolved. James J. Wadsworth never engaged in invective; he never lost his temper. He maintained excellent rapport with the Russians. The fact that negotiations continued for as long as they did, despite deadlock, was as much a tribute to his personal diplomacy as to any other factor.

Even Jerry Wadsworth's good will, however, was not enough to close the gap between American insistence that a nuclear test-ban treaty would be possible only with adequate inspection and the Russian insistence that inspection would stand in the way.

It is possible, however, that one of the earlier specific benefits of Ambassador Wadsworth's patient, constructive negotiating was that the Soviet Union and the United States had been able to agree in October, 1958, on an undeclared moratorium on testing. It was not a formal agreement or a treaty. What happened was that President Eisenhower declared that the United States would abstain from testing so long as the Soviet Union did the same.

As month after month passed without either the United States or the Soviet Union setting off nuclear test explosions, it seemed that a historical advance had been achieved. Premier Khrushchev, on January 14, 1960, hailed the moratorium, declaring that the first nation to resume nuclear tests would "cover itself with shame and . . . will be branded by all the peoples of the world."

On August 30, 1961, however, the Soviet Union broke the moratorium, announcing weapons tests. And on September 5, 1961, the United States announced a new test series of its own.

Not since the beginning of the campaign to end nuclear testing did hopes for sanity among nations drop so sharply.

The successor to Ambassador Wadsworth at the Geneva negotiations in 1961 was Arthur H. Dean. Dean was much more than a man merely trying to carry out an assignment. His sense of purpose may be apparent in this passage drawn from *Test Ban and Disarmament*, his account of the negotiations:

"One afternoon in late 1958 at a reception a young mother said to me that she was afraid to bear any more children because of the contamination of the air by nuclear testing and because of the possible destruction of the world by nuclear weapons. I tried to assure her that nuclear testing would be stopped and nuclear weapons brought under control. She turned and said, 'Well, what are you doing about it? Why are you so sure this problem will be solved? What assurance do you really have to offer?'

"I thought often about her words and continued to study the problem. But what was I really doing to help?

"So when President Kennedy . . . , through his Special Adviser on Disarmament, John J. McCloy, asked me to work with him on the proposed nuclear test-ban treaty and on disarmament, I responded gladly. . . ."

On August 27, 1962, the United States and the United Kingdom submitted alternative draft treaties to the Test Ban Subcommittee of the Eighteen-Nation Disarmament Committee in Geneva.[1] The first, a draft treaty banning tests in all environments, provided for control by an international commission which would supervise national control posts and conduct a limited but unspecified number of on-site inspections each year of unidentified seismic events.

The second proposed treaty would ban tests only in the atmosphere, in outer space, and under water. This limited treaty did not require any international control or inspection, since the Western powers believed that compliance could be monitored by national detection systems.

---

[1] Members of the Committee: Canada, France, Italy, United Kingdom, U.S.A.; Bulgaria, Czechoslovakia, Poland, Rumania, U.S.S.R.; Brazil, Burma, Ethiopia, India, Mexico, Nigeria, Sweden, and U.A.R. The ten members of the original committee, established in September, 1959, are listed first; France refused to participate. Brazil and Mexico, included among the "non-aligned" eight, are members of the Organization of American States.

The Soviets returned to their earlier position that underground tests could be detected at a distance and that no on-site inspections were necessary. They considered Western insistence on on-site inspection a strictly political demand, at best designed to prevent agreement on a treaty, and at worst to facilitate Western espionage.

The negotiations were deadlocked and stayed that way until after the Cuban missile crisis.

One fact that ought to be emphasized here is that the test-ban issue was interwoven with the broad question of American-Soviet relations and the even larger question of world peace. And all the activities, public and private, that had a bearing on these questions, had some bearing on the events leading up to and away from the Cuban crisis.